Ernst Schering Research Foundation Workshop 17
Alzheimer's Disease

Springer-Verlag Berlin Heidelberg GmbH

Ernst Schering Research Foundation
Workshop 17

Alzheimer's Disease

Etiological Mechanisms
and Therapeutic Possibilities

J.D. Turner, K. Beyreuther, F. Theuring
Editors

With 21 Figures and 11 Tables

 Springer

Series Editors: G. Stock and U.-F. Habenicht

ISBN 978-3-662-03250-3

Die Deutsche Bibliothek – CIP-Einheitsaufnahme
Schering-Forschungsgesellschaft <Berlin>:
Ernst Schering Research Foundation Workshop.

ISSN 0947-6075
NE: HST
17. Alzheimer's Disease. – 1996
Alzheimer's disease : etiological mechanisms and therapeutic possibilities ; with 11
tables / J.D. Turner ... ed.

(Ernst Schering Research Foundation Workshop; 17)
ISBN 978-3-662-03250-3   ISBN 978-3-662-03248-0 (eBook)
DOI 10.1007/978-3-662-03248-0
NE: Turner J.D. [Hrsg.]

CIP data applied for

Typesetting: Data conversion by Springer-Verlag

21/3135–5 4 3 2 1 0 – Printed on acid-free paper

Preface

In the Western world Alzheimer's disease is one of the leading causes of senile dementia-induced loss of memory and reason in the elderly. Other major causes include the occurrence of multiple strokes or Parkinson's disease.

The clinical manifestations of Alzheimer's disease have been described in detail and the accuracy of diagnosis has improved. Studies of the pathology have also made substantial progress. The classical hallmarks of the disease, such as senile plaques, neurofibrillary tangles, and neuronal loss have now been investigated in great detail.

The etiology and pathogenesis of Alzheimer's disease are presently not well understood. However, during the last 5 years, considerable research has been devoted toward understanding the basic molecular mechanisms of this multifactorial disease and to developing diagnostic and therapeutic strategies.

The individual chapters in this volume report on the latest developments in these areas. The organizers of the workshop hope that this book will contribute to a better understanding of Alzheimer's disease and the underlying mechanisms.

J.D. Turner
K. Beyreuther
F. Theuring

The participants of the workshop

Table of Contents

List of Editors and Contributors

Editors

J.D. Turner
Schering AG, Neuropsychopharmcology, Müllerstraße 178, 13342 Berlin,
Germany

K. Beyreuther
Center for Molecular Biology, University of Heidelberg,
Im Neuenheimer Feld 280, 69120 Heidelberg, Germany

F. Theuring
Schering AG, Institute of Cellular and Molecular Biology, Müllerstraße 178,
13342 Berlin, Germany

Contributors

O. Almkvist
Karolinska Institutet, Department of Clinical Neuroscience
and Family Medicine, Division of Geriatric Medicine (B84),
Huddinge University Hospital, 14186 Huddinge, Sweden

D. Anthony
Department of Pharmacology, University of Oxford, Mansfield Road,
Oxford OX1 3QT, UK

M.D.Bell
Department of Pharmacology, University of Oxford, Mansfield Road,
Oxford OX1 3QT, UK

K.M. Einhäupl
Department of Neurology, Charité, Humboldt-University,
Schumannstraße 20-21, 10098 Berlin, Germany

J.D. Gearhart
The Johns Hopkins University School of Medicine, Houck 255,
600 N. Wolfe Street, Baltimore, MD 21287, USA

S.D. Ginsberg
Neuropathology Laboratory, The Johns Hopkins University School
of Medicine, 558 Ross Research Building, 720 Rutland Avenue,
Baltimore, MD 21205-2196, USA

C. Haass
Central Institute of Mental Health, University of Heidelberg, J5,
68159 Mannheim, Germany

J. Hardy
Suncoast Alzheimer's Disease Laboratory, Departments of Psychiatry,
Pharmacology, Neurology and Biochemistry, University of South Florida,
Tampa, FL 33613, USA

M. Hutton
Suncoast Alzheimer's Disease Laboratory, Departments of Psychiatry,
Pharmacology, Neurology and Biochemistry, University of South Florida,
Tampa, FL 33613, USA

V. Jelic
Karolinska Institutet, Department of Clinical Neuroscience
and Family Medicine, Division of Geriatric Medicine (B84),
Huddinge University Hospital, 14186 Huddinge, Sweden

C.S. von Koch
Neuropathology Laboratory, The Johns Hopkins University School
of Medicine, 558 Ross Research Building, 720 Rutland Avenue,
Baltimore, MD 21205-2196, USA

B.T. Lamb
Department of Obstetrics and Gynecology, The Johns Hopkins University
School of Medicine, Park B 202, 600 N. Wolfe Street, Baltimore, MD 21287,
USA

L. Lannfelt
Karolinska Institutet, Department of Clinical Neuroscience
and Family Medicine, Division of Geriatric Medicine (B84),
Huddinge University Hospital, 14186 Huddinge, Sweden

M.K. Lee
Neuropathology Laboratory, The Johns Hopkins University School
of Medicine, 558 Ross Research Building, 720 Rutland Avenue,
Baltimore, MD 21205-2196, USA

V.M.-Y. Lee
Department of Pathology and Laboratory Medicine, Division of Anatomic
Pathology, University of Pennsylvania School of Medicine,
Philadelphia, PA 19104-4283, USA

A.C.Y. Lo
Institute of Molecular Biology, The University of Hong Kong,
8 Sassoon Road, Hong Kong

E. Masliah
Department of Neuroscience, University of California, San Diego, Medical
Teaching Building Room 348, 9500 Gilman Drive, La Jolla, CA 92093, USA

M. Mawal-Dewan
Department of Pathology and Laboratory Medicine, Division of Anatomic
Pathology, University of Pennsylvania School of Medicine,
Philadelphia, PA 19104-4283, USA

R.M. Nitsch
Center for Molecular Neurobiology, University of Hamburg,
Martinistraße 52, 20246 Hamburg, Germany

A. Nordberg
Karolinska Institutet, Department of Clinical Neuroscience
and Family Medicine, Division of Geriatric Medicine (B84),
Huddinge University Hospital, 14186 Huddinge, Sweden

G. Perry
Institute of Pathology, Case Western Reserve University,
2085 Adelbert Road, Cleveland, OH 44106, USA

V.H. Perry
Department of Pharmacology, University of Oxford, Mansfield Road,
Oxford OX1 3QT, UK

L.H.T. Van der Ploeg
Merck & Co., Inc., Department of Genetics and Molecular Biology,
P.O. Box 2000, RYBOY-255, Rahway, NJ 07065-0900, USA

D.L. Price
Neuropathology Laboratory, The Johns Hopkins University School
of Medicine, 558 Ross Research Building, 720 Rutland Avenue,
Baltimore, MD 21205-2196, USA

A.J.I. Roskams
Neuropathology Laboratory, The Johns Hopkins University School
of Medicine, 558 Ross Research Building, 720 Rutland Avenue,
Baltimore, MD 21205-2196, USA

N.J. Rothwell
School of Biological Sciences, University of Manchester,
Oxford Road, 1.124 Stopford Building, Manchester, M13 9PT, UK

J.G. Schulz
Department of Neurology, Charité, Humboldt-University,
Schumannstraße 20-21, 10098 Berlin, Germany

M. Shigeta
Karolinska Institutet, Department of Clinical Neuroscience
and Family Medicine, Division of Geriatric Medicine (B84),
Huddinge University Hospital, 14186 Huddinge, Sweden

S.S. Sisodia
Neuropathology Laboratory, The Johns Hopkins University School
of Medicine, 558 Ross Research Building, 720 Rutland Avenue,
Baltimore, MD 21205-2196, USA

H.H. Slunt
Neuropathology Laboratory, The Johns Hopkins University School
of Medicine, 558 Ross Research Building, 720 Rutland Avenue,
Baltimore, MD 21205-2196, USA

M.A. Smith
Institute of Pathology, Case Western Reserve University,
2085 Adelbert Road, Cleveland, OH 44106, USA

E.R. Stadtman
Laboratory of Biochemistry, National Heart, Lung, and Blood Institute,
National Institutes of Health, Building 3, Room 222, Bethesda, MD 20892,
USA

G. Thinakaran
Neuropathology Laboratory, The Johns Hopkins University School
of Medicine, 558 Ross Research Building, 720 Rutland Avenue,
Baltimore, MD 21205-2196, USA

J.Q. Trojanowski
Department of Pathology and Laboratory Medicine, Division of Anatomic
Pathology, University of Pennsylvania School of Medicine,
Philadelphia, PA 19104-4283, USA

L.-O. Wahlund
Karolinska Institutet, Department of Clinical Neuroscience
and Family Medicine, Division of Geriatric Medicine (B84),
Huddinge University Hospital, 14186 Huddinge, Sweden

B. Winblad
Karolinska Institutet, Department of Clinical Neuroscience
and Family Medicine, Division of Geriatric Medicine (B84),
Huddinge University Hospital, 14186 Huddinge, Sweden

H. Zheng
Merck Research Laboratories, Inc., Department of Genetics
and Molecular Biology, 126 East Lincoln Avenue, Rahway, NJ 07065, USA

1 Alzheimer's Disease With and Without Familial Aggregation: A Case for Phenotypical Similarity

O. Almkvist, V. Jelic, L. Lannfelt, A. Nordberg, M. Shigeta, L.-O. Wahlund, and B. Winblad

1.1 General Introduction

Alzheimer's disease (AD) is a neurodegenerative disease affecting the brain, with devastating implications for the afflicted individual due to the gradual loss of cognitive functions and independence in living, which finally result in death. AD generates an ever-growing expenditure for society from the need to care for the individuals who cannot live independently and patients who require institutionalized care.

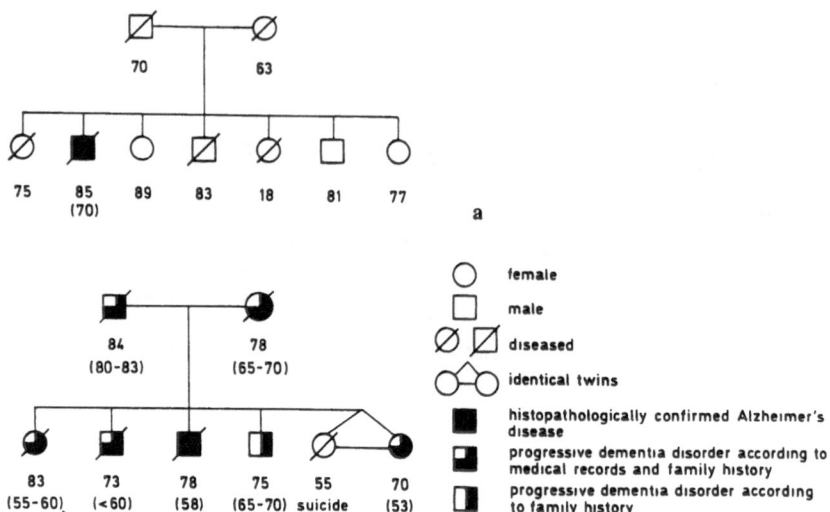

Fig. 1a,b. Typical pedigrees of simplex (**a**) and multiplex Alzheimer's disease AD (**b**) showing age at death and age at onset (within parentheses)

AD may show familial aggregation and to some extent the cause of this aggregation is known (familial AD, FAD). Early-onset AD has three known genetic causes. These are mutations in the amyloid precursor protein (APP) on chromosome 21 (AD$_{APP}$), mutations in the presenilin-1 protein (PS-1) on chromosome 14 (AD$_{PS-1}$), and mutations in the second seven-transmembrane protein on chromosome 1 (AD$_{STM2}$). It is known that there is a strong association between apolipoprotein E (Apo E) located on chromosome 19 and the onset age of AD. However, familial aggregation of AD may also be coupled with unknown causes, here termed multiplex AD (MAD). The total prevalence of AD with familial aggregation has been estimated to range from 10% to 75% (Appel 1981; Breitner et al. 1988).

AD may also occur without any familial aggregation, i.e., sporadic or simplex AD (SAD). SAD has been associated with a number of risk factors. Down's syndrome, increasing age, and head trauma have been confirmed as risk factors for AD from epidemiological research (Jorm 1990). Toxin exposure, female gender, low level of education, immunological factors, as well as other demographic and geographic variables have been suggested but not confirmed (Fig. 1; Jorm 1990).

In this paper, individuals with AD will be analyzed with respect to their former history of AD. The analyses include neuropathological (study I) and clinical features, EEG findings, Apo E genotype, positron emission tomography (PET) imaging of the brain, and cognitive functions (study II). The general purpose was to explore whether SAD and MAD could be differentiated based upon neuropathological changes and/or clinical characteristics.

1.2 Study I: Neuropathology and MAD Versus SAD

1.2.1 Methods

Two groups of subjects were studied, 12 patients with MAD and 29 patients with SAD. All patient had been cared for at the Psychogeriatric Unit at Umeå University hospital between 1978–1982. A brain autopsy was performed on all patients according to Alafuzoff et al. (1987).

1.2.2 Results

Some of the clinical characteristics are presented in Table 1. There was no statistical difference between MAD and SAD in clinical characteristics such as sex distribution, birth order, onset age, duration of the disease, or mean maternal/paternal age at birth, using Student's t-test.

Table 1. Subject characteristics of study I

	MAD (n=12)	SAD (n=29)
Gender (female/male)	4/8	11/18
Age at onset (Mean±SD)	69.0±10.4	73.0±7.8
Duration (years)	6.3±4.7	5.3±3.7
Maternal age at birth of the diseased	30.9±5.5	31.0±7.0
Paternal age at birth of the diseased	35.2±7.2	35.0±7.2

MAD, Multiplex Alzheimer's disease; *SAD*, simplex Alzheimer's disease.

Table 2. Family characteristics of MAD and SAD

	MAD ($n=12$)	SAD ($n=29$)
Number of diseased in second generation	33	29
Probands/secondary	12/21	29/0
Gender (female/male)	17/16	11/18
Age at onset (Mean±SD)	67.0±10.4	73.0±7.8 **
Duration (years)	7.9±6.4	5.3±3.7
Onset <65	11.8±7.2	6.3±4.4 **
Onset >65	4.1±2.5	5.2±3.3
Maternal age at birth of the diseased	31.0±7.0	30.8±6.3
Paternal age at birth of the diseased	35.0±7.2	35.7±6.2

MAD, Multiplex Alzheimer's disease; *SAD,* simplex Alzheimer's disease.
*=$p<.05$; **=$p<.01$

Table 3. Amount and distribution of neuropathological changes for MAD and SAD

	MAD ($n=12$)	SAD ($n=2$)
Senile/neuritic plaques		
Frontal cortex	10.3±7.7	8.2±6.4
Hippocampus	12.8±8.5	10.0±4.6
Neurofibrillary tangles		
Frontal cortex	3.3±2.4	5.3±5.4*
Hippocampus	14.4±15.4	19.1±12.2

MAD, Multiplex Alzheimer's disease; *SAD,* simplex Alzheimer's disease.
*=$p<.05$

The family characteristics of MAD and SAD are presented in Table 2. The age of onset was significantly earlier in MAD than in SAD. The duration of the disease was significantly longer in MAD than in SAD. There was no difference in maternal/paternal age at birth of the diseased subject.

A regression analysis showed a significant correlation ($r=0.86$) between the age of onset for study subjects and secondary cases (i.e., cases of dementia within the study subject's family among parents, sibling and/or children; see Fig. 2.

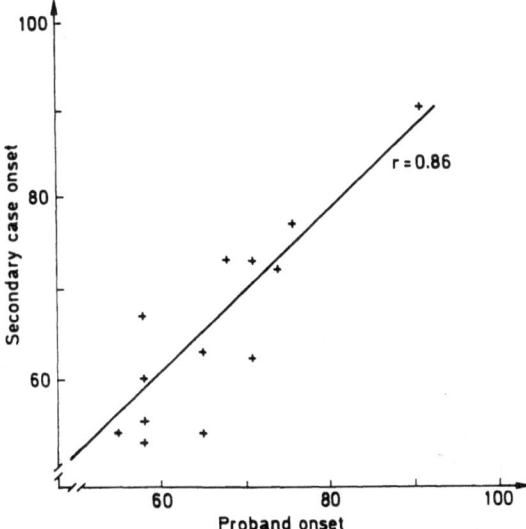

Fig. 2. Correlation between onset age for study subjects and secondary cases (cases of dementia within the study subject's family belonging to the generation of parents, siblings, and/or children)

The amount of senile/neuritic plaques and neurofibrillary tangles (NFT) were compared in frontal cortex and hippocampus; no changes were revealed between SAD and MAD. The NFTs were more abundant in SAD than in MAD and this difference reached statistical significance in the frontal regions, but not in the hippocampus (Table 3).

The finding that the age of onset for secondary cases correlated with the age of onset for the study subject is in agreement with previous studies, showing a tendency for onset age to be a family trait (Axelman et al. 1994; Breitner et al. 1988).

1.3 Study II: Clinical and Functional Features and MAD Versus SAD

1.3.1 Methods

For a subset of patients examined for suspected dementia at the Department of Geriatric Medicine, Huddinge University Hospital between 1992 and 1994, there was an investigation of brain glucose metabolism using [18F]fluoro-deoxy-glucose ([18F]FDG) and PET in order to increase the diagnostic accuracy of dementia disorders. All PET studies were performed at the PET Center, Uppsala University. The investigation was carried out when subjects were at rest, using [18F]FDG which was injected as a tracer intravenously and simultaneously with the start of the PET camera (PC 4096 GEMS AB Scanditronix, Uppsala, Sweden). Arterial blood samples were drawn from a catheter in the left radial artery and analyzed for plasma radioactivity. The rate of [18F]FDG influx into the hexokinase state was analyzed according to the method of Patlak and coworkers (Patlak et al. 1983). The glucose utilization was calculated by multiplying this rate by the plasma concentration of glucose and dividing it by a lumped constant of 0.418 to correct for differences in the utilization of FDG as compared to glucose. After injection, the camera recorded images continuously with an increasing frame time from 1 to 60 min postinjection. Both the right and the left sides were analyzed in seven areas: sensorimotor cortex, visual cortex, temporoparietal association cortex, frontal cortex, thalamus, putamen, and cerebellum. In order to minimize interindividual variation, the ratio between glucose consumption in the region of interest was divided by corresponding values for the sensorimotor cortex.

The diagnosis of dementia followed the criteria of DSM-III-R (American Psychiatric Association 1987). AD was diagnosed according to NINCDS-ADRDA criteria (McKhann et al. 1984) and the patients divided into two groups; there were 18 patients with MAD and 21 with SAD who participated in the present study. The criterion for MAD was the existence of at least one first-degree relative (i.e., parents or siblings) with AD. The criterion for SAD was a negative history of AD in all first-degree relatives (i.e., parents or siblings).

All AD patients underwent a number of other examinations. DNA was prepared from peripheral blood by standard procedures, and Apo E genotypes were analyzed using the method of Wenham et al. (1991).

EEGs were obtained with a 16-channel Siemens recorder using the international 10–20 system of electrode placement (time constant: 0.3 s; low-pass filter: 30 Hz). The EEG was recorded in a resting, awake condition with closed eyes. Part of the EEG was digitized at a sampling rate of 128 Hz (Brain Atlas, Bio-Logic Systems), 20 2-s periods of artifact-free recordings were selected and a frequency analysis was performed using a FFT algorithm with a Hanning window. Twelve scalp bipolar derivations were chosen in the left and right parietal (C3–P3, C4–P4), occipital (T5–O1, T6–O2), and temporal regions (T3–T5, T4–T6). The EEG parameters chosen were the relative power in four frequency bands: delta (2–4 Hz), theta (4–8 Hz), alpha (8–13 Hz), and beta (13–20 Hz) and the mean frequency averaged across the spectrum. Relative power values were transformed using the $\log (x/1-x)^{1/2}$ to normalize the distribution of data.

The neuropsychological assessment was performed by an experienced neuropsychologist and included general cognition by means of the Full Scale Intelligence Quotient of the Wechsler Adult Intelligence Scale-Revised (FSIQ; Wechsler 1981), verbal abilities by means of abstraction (Similarities; Wechsler 1981), visuospatial functions by means of constructional praxis (Block Design; Wechsler 1981), verbal short-term memory (Digit Span; Wechsler 1981), verbal episodic memory (Stockholm Gerontology Research Center, Free Recall; SGRC: FR; Bäckman and Forsell 1994 and Rey Auditory Verbal Learning Test; Lezak 1983), and attention by means of tracking speed (Digit Symbol; Wechsler 1981 and Trail Making, part A; Lezak 1983).

1.3.2 Clinical Results

The subject characteristics are presented in Table 4. For the demographic variables (age, sex, and education), there were no reliable differences between the groups of MAD and SAD patients according to a one-way ANOVA ($Fs<1$). Furthermore, there were no reliable differences between groups of MAD and SAD patients (one-way ANOVA, $Fs<1$) in duration of the disease or in general cognitive level as indica-

Table 4. Subject characteristics of study II

	MAD (n=18)	SAD (n=21)
Age (mean±SD)	59.2±9.2	59.6±6.5
Range	41–79	50–72
Gender (female/male)	9/9	13/8
Education in years (mean±SD)	10.8±3.5	10.3±3.9
Range	7–22	6–19
Age of onset (mean±SD)	55.6±9.0	57.1±6.1
Range	37–78	49–70
Duration in years (mean±SD)	2.8±2.5	2.4±1.5
Range	0–9	1–6
MMSE (mean±SD)	22.2±4.1	21.8±4.4
Range	12–27	15–30

None of the values were significant for either MAD or SAD.
MAD, Multiplex Alzheimer's disease; *SAD*, simplex Alzheimer's disease; *MMSE*, Mini-Mental Satate Examination.

ted by the Mini-Mental Examination Score (MMSE; Folstein et al. 1975).

1.3.3 Apo E Results

The distribution of Apo E alleles (see Table 5) was not significantly different for MAD and SAD (chi-squared test, $p<.1$), although it differed from expectation-based prevalence rates in healthy individuals within the same age range (Eggertsen et al. 1993). There was a lower predominance of ε3/ε3 alleles in our two AD groups (33% and 35%, respectively) compared to that reported in normal elderly individuals (59%, see Eggertsen et al. 1993). At the same time, the frequency of at least one ε4 allele is similarly elevated in our two patient groups (50% and 45%, respectively) compared to normal elderly individuals (37%, see Eggertsen et al. 1993; Lannfelt et al. 1995).

Table 5. Distribution of apolipoprotein E alleles in MAD and SAD

Apolipoprotein E alleles	MAD (n=18)	SAD (n=20)
ε2/ε2	0	0
ε2/ε3	2	4
ε3/ε3	6	7
ε2/ε4	1	0
ε3/ε4	4	6
ε4/ε4	5	3

MAD, Multiplex Alzheimer's disease; *SAD*, simplex Alzheimer's disease.

1.3.4 EEG Results

The most interesting results from the EEG examination are presented in Table 6, showing those that emanate from the parietal cortex of the left (C3–P3) and right hemisphere (C4–P4). The associated cortical areas are known to have a large proportion of degenerative changes early in the disease course. However, no differences between groups of MAD

Table 6. Results from EEG analyses in MAD and SAD showing mean frequency and relative power in four frequency bands for right (C4–P4) and left (C3–P3) parietal regions (mean±SD)

	MAD (n=15)	SAD (n=19)
Mean frequency (Hz)	9.41±1.37	9.48±1.18
Mean relative power		
alpha (C3–P3)	−0.38±0.51	−.25±0.74
alpha (C4–P4)	−0.26±0.62	−.14±0.83
beta (C3–P3)	−1.43±0.85	−1.36±0.84
beta (C4–P4)	−1.54±0.75	−1.32±0.80
delta (C3–P3)	−1.80±0.72	−2.00±0.86
delta (C4–P4)	−1.71±0.80	−2.04±0.74
theta (C3–P3)	−0.44±0.70	−0.69±0.73
theta (C4–P4)	−0.50±0.78	−0.76±0.87

None of the values were significant for either MAD or SAD.
MAD, Multiplex Alzheimer's disease; *SAD*, simplex Alzheimer's disease.

and SAD patients were observed in the mean frequency across all the derivations or in any single derivation. The different derivations demonstrated a varying mean frequency, showing the lowest frequency for the parietal left (C3–P3) and right (C4–P4) derivations. However, both AD groups showed the same pattern of results. There were no statistically significant results on relative power for any of the leads in any of the frequency bands according to one-way ANOVA ($Fs<1$).

1.3.5 PET Results

The result from the PET examinations are presented in Table 7. There was no statistical difference between MAD and SAD in normalized values of glucose metabolism in any of the regions studied (one-way ANOVAs, $Fs<1$). However, the typical bilateral relative hypometabolism (approximately 75% of the sensorimotor cortex) observed in the temporoparietal cortex was similar for both AD groups. Second, some reduction in brain glucose metabolism was observed both in the frontal

Table 7. Results of positron emission tomography for different brain regions in MAD and SAD showing normalized values of glucose metabolism (regional absolute value divided by the value of sensorimotor cortex)

Region	MAD (n=17)	SAD (n=21)
Visual cortex, right	1.24±0.30	1.20±0.21
Visual cortex, left	1.23±0.29	1.21±0.23
Temporo-parietal association cortex, right	0.80±0.10	0.74±0.19
Temporo-parietal association cortex, left	0.75±0.15	0.67±0.16
Frontal cortex, right	0.93±0.16	0.94±0.11
Frontal cortex, left	0.91±0.20	0.91±0.11
Thalamus, right	1.18±0.11	1.17±0.17
Thalamus, left	1.19±0.14	1.18±0.14
Putamen, right	1.29±0.15	1.25±0.16
Putamen, left	1.28±0.17	1.28±0.14
Cerebellum, right	0.88±0.19	0.91±0.12
Cerebellum left	0.91±0.15	0.95±0.11

None of the values were significant for either MAD or SAD.

cortex and the cerebellum (approximately 90% of the sensorimotor cortex), although the reduction was similar for both MAD and SAD. In some regions, there was a clear assymmetry between the left and right hemisphere, but this did not differ between the two AD types.

1.3.6 Neuropsychological Results

The results from neuropsychological testing of MAD and SAD patients are presented in Table 8. The two groups did not differ in mean level of performance according to one-way ANOVAs in general cognition, verbal abstraction, constructional praxis, short-term memory, episodic memory, or attention (Fs<1). Although there was no difference between AD groups, the results did show a marked impairment compared to the normal performance for healthy elderly individuals in most tests (more than 2 standard deviations below the mean).

The neuropsychological test results strongly correlated with brain glucose metabolism in the temporoparietal cortex, less strongly with the frontal areas, and only slightly with other brain regions.

Table 8. Neuropsychological test results (mean±SD) in MAD and SAD

Function (test)	MAD (n=18)	SAD (n=21)
Cognition (FSIQ)	82.1±14.5	77.6±14.8
Verbal abstraction (Similarities)	13.6±8.2	11.8±6.9
Visuospatial ability (Block Design)	12.1±10.1	7.3±9.3
Short-term memory (Digit Span)	9.8±3.8	9.0±3.9
Episodic memory (SGRC: FR)	2.9±1.6	2.3±2.4
Episodic memory (Rey AVLT)	23.7±9.5	22.9±11.2
Attention (Digit Symbol)	15.0±9.2	12.9±16.8
Attention (TMT: part A)[a]	110±55	136±89

None of the values were significant for either MAD or SAD.

FSIQ, Full Scale Intelligence Quotient; *SGRC,* Stockholm Gerontology Research Center; *FR,* Free Recall; *AVLT,* Auditory Verbal Learning Test; *TMT,* Trail Making Test

[a] The lower the result, the better, because performance is measured in seconds.

1.4 General Discussion

The purpose of the present study was to investigate whether a differential pattern in clinical features or neuropathological changes might exist between MAD and SAD from hospital-based samples of patients. This study was based upon two individual studies, one in which neuropathological findings were descibed and one in which clinical features were presented.

In study I, a comparison of histopathological changes in the SAD and MAD patients revealed no major differences in the distribution or number of degenerative changes such as senile/neuritic plaques or NFTs. Our findings are in agreement with previous attempts to differentiate etiological subgroups of AD. Thus, it was shown in a study by Katzman (1976) that the neuropathological characteristics of SAD could not be distinguished from those of FAD. As far as we know, no study has been performed comparing MAD and SAD or FAD and MAD.

In study II, the general pattern of results did not reveal any obvious differences between MAD and SAD in clinical features, examinations of Apo E allele distribution, EEG patterns, brain glucose metabolism, or neuropsychological tests. The main conclusion of our study is negative, demonstrating a similarity of clinical expressions in MAD and SAD in spite of the difference in supposed etiology of the disease.

In previous studies, the pattern of results does not differentiate SAD and FAD according to clinical data, neuroimaging results, or neuropsychological data (Duara et al. 1993; Swearer et al. 1992). Duara and coworkers (1993) found an earlier onset and a longer duration of the disease in FAD, but failed to find any phenotypic differences between FAD and SAD in terms of cognitive functions (except for a relatively worse language performance in SAD) or results of MR imaging or PET examination of brain glucose metabolism. Swearer and associates (1992) found a similar pattern of neuropsychological impairment in groups of FAD and SAD patients matched for global disease severity. It was also shown that there was no interaction between type of AD and age, indicating that early and late onset does not influence the pattern of cognitive impairment (Swearer et al. 1992). Previous arguments that early aphasic or apraxic deficits are typical for FAD (Breitner and Folstein 1984; Chui et al. 1985) may not be reliable due to confounding effects of age, education, and severity of global cognitive decline (Swearer et al. 1992).

Furthermore, the rate of cognitive deterioration as measured by the Blessed Dementia Score (Blessed et al. 1968) was found to be unrelated to the risk of developing a genetically determined type of AD (Farrer et al. 1995). Similarly, functional decline as assessed by Weintraub (1986) has been found to be unrelated to the risk of developing a genetically determined type of AD (Farrer et al. 1995).

With regard to Apo E genotype frequency, contradictory results have been reported. Some studies report negative results (Lannfelt et al. 1995), whereas other studies have reported a lower ε4 allele frequency in SAD than in FAD (see, for example, vanDuijn et al. 1994).

There are unpublished data concerning the molecular processes involved in AD which show similar mean values and ranges of CSF τ proteins, a metabolite of NFT, in FAD and SAD (O. Almkvist et al., unpublished data). In contrast, there are also single case autopsy observations of AD_{APP21} showing a lower amount of both senile plaques and NFTs than in AD_{PP14} and a lower amont of tau staining in AD_{APP21} than in AD_{APP14} (Bogdanovic, personal communication). The answer to the question of whether these observations are specific for individuals or specific for genotypes must await further studies.

To summarize, the present study confirms the results from previous studies, indicating that there is no specificity in phenotypes related to sporadic AD and AD with familial aggregation. In this study we were unable to differentiate MAD and SAD based upon clinical methods and neuropathological examinations. Since the clinical manifestations of MAD and SAD are the same, MAD and SAD have to be considered as one disease entity until contradictory data are presented.

Acknowledgment. This research was supported by a grants from the Swedish Council for Research in the Humanities and Social Sciences, the Gamla Tjäna-rinnor Foundation, the Swedish Council for Social Research, the Einar Belvén Foundation, the Torsten and Ragnar Söderbergs Foundation, the Municipal Pension Institute, and SHMF-90. Comments on the manuscript from Philippa Loyd are gratefully acknowledged.

References

Alafuzoff I, Iqbal K, Friden H, Adolfsson R, Winblad B (1987) Histopathological criteria for progressive dementia disorders: clinical-pathological correlation and classification by multivariate data analysis. Acta Neuropathol (Berl) 74:209–225

Alafuzoff I, Almqvist E, Adolfsson R, Lake S, Wallace W, Greenberg DA, Winblad B (1994) A comparison of multiplex and simplex families with Alzheimer's disease/senile dementia of the Alzheimer type within a well defined population. J Neural Transm Park Dis Dement Sect 7:61–72

American Psychiatric Association (1987) Diagnostic and statistical manual of mental disorders, 3rd edn. APA, Washington

Appel SH (1981) A unifying hypothesis for the cause of amytrophic lateral sclerosis, parkinsonism and Alzheimer's disease. Ann Neurol 10:499–505

Axelman K, Basun H, Wahlund L-O, Lannfelt L (1994) A clinical and genealogical investigation of a large Swedish Alzheimer family with a codon 670/671 amyloid precursor protein mutation. Arch Neurol 51:1193–1197

Bäckman L, Forsell Y (1994) Episodic memory functioning in a community-based sample of very old adults with major depression: utilization of cognitive support. J Abnorm Psychol 103:361–370

Blessed G, Tomlinson BE, Roth M (1968) The association between quantitative measures of dementia and the senile changes in the cerebral grey matter of elderly subjects. Br J Psychol 225:797–811

Breitner JCS, Folstein MF (1984) Familial Alzheimer's dementia: a prevalent disorder with specific clinical features. Psychol Med 14:63–80

Breitner JCS, Silverman JM, Molls RC, Davis KL (1988) Familial aggregation in Alzheimer's disease: comparison of risk among relatives of early- and late-onset cases, and among male and femal relatives in successive generations. Neurology 38:207–212

Chui HC, Teng EL, Henderson VW, Moy AC (1985) Clinical subtypes of dementia of the Alzheimer type. Neurology 35:1544–1550

Duara R, Lopez-Alberola RF, Barker WW, Loewenstein DA, Zatinsky M, Eisdorfer CE, Weinberg GB (1993) A comparison of familial and sporadic Alzheimer's disease. Neurology 43:1377–1384

Eggertsen G, Tegelman R, Ericsson S, Berglund L (1993) Apolipoprotein E polymorphism in a healthy Swedish population: variation of allele frequency with age and relation to serum lipid concentrations. Clin Chem 39:2125–2129

Farrer LA, Cupples LA, vanDuijn CM, Connor-Lacke L, Kiely DK, Growdon JH (1995) Rate of progression of Alzheimer's disease is associated with genetic risk. Arch Neurol 52:918–923

Folstein MF, Folstein SE McHugh PR (1975) "Mini-Mental State Examination:" a practical method for grading the cognitive status of the patient for the clinician. J Psychiatric Res 12:189–198

Jorm A (1990) The epidemiology of Alzheimer's disease and related disorders. Chapman and Hall, London

Katzman R (1976) The prevalence and malignancy of Alzheimer' s disease. Arch Neurol 33:217–218

Lannfelt L, Pedersen NL, Lilius L, Axelman K, Johansson K, Viitanen M, Gatz M (1995) Apolipoprotein e4 allele in Swedish twins and siblings with Alzheimer's disease. Alzheimer Dis Assoc Disord 9: 166–169

Lezak M (1983) Neuropsychological assessment, 2nd edn. Oxford Univerity Press, New York

McKhann G, Drachman D, Folstein M, Katzman R, Price D, Stadlan EM (1984) Clinical diagnosis of Alzheimer's disease: report of the NINCDS-ADRDA work group under the auspices of the department of health and human service task force on Alzheimer's disease. Neurology 34:939–944

Patlak CS, Blasberg JD, Fenstermacher JD (1983) Graphical evaluation of blood to brain transfer constants from multiple tissue uptake. J Cereb Blood Flow Metab 3:1–7

Swearer JM, O'Donnell BF, Drachmann DA, Woodward BM (1992) Neuropsychological features of familial Alzheimer's diseae. Ann Neurol 32:687–694

vanDuijn CM, Van Broeckhoven C, Hardy J et al (1991) Evidence for allelic heterogeneity in familial early-onset Alzheimer's disease. Br J Psychiatry 158:471- 474

vanDuijn CM, deKnijff P, Cruts M et al (1994) Apolipoprotein e4 allele in a population-based study of early-onset Alzheimer's disease. Nature Genet 7:74–78

Wechsler D (1981) Manual for the Wechsler Adult Intelligence Scale-Revised. Psychological Corporation, New York

Weintraub S (1986) The record of independent living: an informant-completed measure of activities of daily living and behavior in elderly patients with cognitive impairment. Am J Alzheimer Care 1:35–39

Wenham PR, Price WH, Blundell G (1991) Apolipoprotein E genotyping by one-stage PCR. Lancet 337:1158–1159

2 The Vascular Dementias and Cerebrovascular Involvement in Alzheimer's Disease

J.G. Schulz and K.M. Einhäupl

2.1 Introduction

The blood stream is a trade route between the brain and the rest of the body. It provides various nutrients and clears waste products, receives and delivers signals, may be blocked or disrupted, and exposes the brain to harmful and unknown external agents. How disturbed vascular trafficking in the brain may end in a syndrome that presents clinically as dementia may be categorized in the following way:

1. Global cerebral ischemia due to cardiogenic dysfunction (watershed infarcts)
2. Focal cerebral ischemia due to vascular occlusion (multi-infarct dementia, MID; strategic infarct)
3. Chronic hypoperfusion due to functional changes of the cerebral microvasculature (Binswanger's disease, Alzheimer's disease, AD?)
4. Cerebral hemorrhage due to focal weakness of the cerebrovasculature (intracerebral hemorrhage, amyloid angiopathy)

In acute ischemic processes with complete infarction the clinical picture directly depends upon how large a lesion is and where it is located, because the brain tissue involved is simply lost for further use. This occurs instantaneously and can be detected by neuroimaging. In contrast, whether and how chronic ischemic processes or incomplete infarction may affect the brain may depend upon further unknown variables. They are not necessarily detected by neuroimaging and do not coincide with onset of clinical symptoms. As a consequence, it is difficult to attribute slowly progressing dementia to underlying vascular causes. Therefore such causes may be underestimated and remain to be elucidated.

Whether cerebrovascular changes play a major role or none at all for the neuronal degeneration and resulting dementia in AD can certainly not be answered at this point. However, the idea of a "vascular component" seems tangible. This chapter will focus (1) on proteins causing vascular and parenchymal cerebral amyloidoses (β-amyloid, βA; cystatin C; prion protein) and on apolipoprotein E (apoE) which is associated with both vascular and brain parenchymal diseases with or without amyloid; (2) on a possible role for ischemic events in AD based on morphological findings in cerebral white matter, capillary changes, and β-amyloid precursor protein (βAPP) expression studies; and (3) on capillaries, platelets, and the blood–brain barrier (BBB) in AD.

```
           RRR            LLLLLLLLLLLL
    N——4——-2——Thr————————————C
              HHHHHHHββββββ
```

Fig. 1. ApoE. *R*, receptor-binding domain; *L*, lipoprotein-binding determinant; *Thr*, thrombin cleavage-site; *2/4*, point mutations; *H*, heparin binding-site; β, β-protein-binding site

2.2 Proteins with a Dual Role: ApoE and Amyloid Proteins Are Associated with Both Vascular and Parenchymal Disease

2.2.1 Apolipoprotein E

The identification of allele ε4 of apolipoprotein E gene (APOE4) as a risk factor in AD changed our understanding of the pathophysiological processes in this and other diseases of the CNS. The human APOE gene on chromosome 19 encodes for a polymorphic 299 amino acid variably sialylated glycoprotein. Two independently folded domains can be separated by thrombin cleavage: the N-terminal part contains a receptor-binding domain (134–150), the C-terminal part (216–299) contains lipoprotein-binding determinants. A heparin binding site is located between residues 202 and 243, a β-peptide (βP) binding site between residues 244 and 272 (see Fig. 1). ApoE3 is the dominant human isoform (more than 75% in humans), from which apoE4 and apoE2 differ by a single amino acid (112 Cys→Arg and 158 Arg→Cys, respectively; for review see Weisgraber et al. 1994).

Systemically, apoE is found in chylomicrons and very low, intermediate, and high density lipoprotein (HDL) but not low density lipoproteins (LDL) and mediates their cellular uptake and degradation. ApoE protects against atherosclerosis as shown in apoE-deficient mice that develop typical lesions and hypercholesterolemia (Zhang et al. 1992; Plump et al.1992). A central role of apoE in systemic lipid and cholesterol metabolism has been suggested (for review see Mahley 1988).

Table 1. Increased APOE4 frequencies in vascular and brain parenchymal diseases

Disease	ε4 allele frequency	Reference
Vascular disease		
Atherosclerosis	0.21	Davignon et al.1988
CAD	0.26	van Bockxmeer and Mamotte 1992
Type V hyperlipidemia	0.25	Ghiselli et al. 1982
CAA	0.44	Greenberg et al. 1995
VD	0.45	Frisoni et al. 1994
MID	0.21	Noguchi et al. 1993
Parenchymal disease		
Late-onset familial AD	0.42	Saunders et al. 1993
Sporadic AD	0.40	Saunders et al. 1993
CJD	0.31	Amouyel et al. 1994
Pick's disease	0.38	Schneider et al. 1995
CBD	0.33	Schneider et al. 1995
PSNP	0.25	Schneider et al. 1995
Lewy body disease	0.35	St. Clair et al. 1994
Controls	0.09–0.18	

CAD, coronary artery disease; *CAA*, cerebral amyloid angiopathy; *VD*, vascular dementia; *MID*, multi-infarct dementia; *AD*, Alzheimer's disease; *CJD*, Creutzfeld-Jakob disease; *CBD*, corticobasal degeneration; *PSNP*, progressive supranuclear palsy.

2.2.1.1 APOE as a Risk Factor

APOE4 has been shown to be a risk factor for vascular diseases (see Table 1), associated with changes in lipoprotein metabolism such as enhanced intestinal cholesterol absorption, higher plasma cholesterol and LDL and higher systolic blood pressure. More recently, higher APOE4 frequencies have been detected in Caucasians and African-Americans with AD or with certain other neurodegenerative disorders (see Table 1). Patients with Huntington's disease, amyotrophic lateral sclerosis, Parkinson's disease, chromosome 14-linked early-onset familial AD (FAD) or black Africans with AD do not have increased APOE4 frequencies (Saunders et al. 1993; Osuntokun et al. 1995; AD Collaborative Group 1993). For vascular dementias (VD) and Creutzfeldt-Jakob disease (CJD) conflicting data have been reported (Roses et al. 1995; Amouyel et al. 1994; Frisoni et al. 1994).

2.2.1.2 ApoE in the Nervous System

The nervous system synthesizes some, but not all, lipoproteins and their receptors, indicating at least partially independent brain lipid and cholesterol metabolism. The CNS is the second major site of apoE mRNA expression after the liver and apoE is the most abundant CSF apolipoprotein (for review, see Weisgraber et al. 1994). In the brain, only astrocytes and vascular smooth muscle cells synthesize and secrete apoE (Majack et al. 1988; Boyles et al. 1985). Astrocytes contain apoE in the perinuclear region and along the processes that end on basement membranes of blood vessels or below the pia mater. Cortical neurons have been shown to contain apoE in their cytosol but not to synthesize apoE (Han et al 1994). In the peripheral nervous system (PNS), apoE is synthesized in the nonmyelinating glia.

ApoE upregulation is observed in CNS astrocytes after 10 min of global ischemia, after entorhinal cortex injury, and in oligodendrocytes after optic nerve injury, but no mRNA is detected in neurons or macrophages (Hall et al. 1995; Stoll et al. 1989). Increased levels of apoE in CSF are found in patients with inflammatory CNS diseases (Carlsson et al. 1991). In the PNS, apoE can be induced in macrophages after nerve injury (Stoll and Müller 1986; Boyles et al. 1989). These cells accumulate apoE and apoA-I, cholesterol and other membrane components from distal to the injury site. A return to baseline is seen after 8 weeks, when regeneration is complete. No accumulation of apoE is seen in nonregenerating peripheral nerves.

These data support the hypothesis that lipid and cholesterol metabolism in the nervous system are regulated independently from the liver. Further, it suggests that enhanced production of apoE in the nervous system may play a role in local lipid and cholesterol transport. However, peripheral nerve regeneration is possible in apoE-deficient mice, the regenerating nerves being morphologically similar to the ones in normal mice, indicating that apoE might be replaceable in this process (Popko et al. 1993).

2.2.1.3 ApoE in AD

In AD, apoE is found within extracellularly located vascular and parenchymal preamyloid and amyloid deposits of βP, as well as in extracellular and rare intracellular neurofibrillary tangles (NFT); apoE mRNA is
increased in AD (Yamada et al. 1995; Namba et al. 1991). ApoE4 is not
only associated with an increased risk of developing late-onset FAD and
sporadic AD but also with increased vascular and parenchymal βP
deposition in AD and after head trauma (which is a risk factor for AD in
itself) (Roberts et al. 1991). However, non-ε4 allele carriers can have
AD with NFT and βA deposits and not all ε4 allele carriers develop AD,
suggesting that ε4 allele is just a trigger but not the cause of AD, in
agreement with the observation that AD age of onset is shifted APOE4
dose dependently to an earlier time. No increase in APOE4 frequencies
of AD patients was found in black Africans in contrast to African-
Americans. Additionally, there is considerably less cerebral βA deposition and a lower prevalence of dementia in black Africans living in
Africa than in African-Americans living in industrialized countries,
whereas the frequency of APOE4 is roughly the same in both control
populations. The fact that APOE4 is also a risk factor for cerebral
amyloid angiopathy (CAA), which is found in most AD cases, might
imply that CAA and AD have common pathogenetic features or that
CAA is not necessarily part of AD but rather that overlap of these two
conditions is coincidental and merely due to the shared risk factor
APOE4 (Greenberg et al. 1995).

The involvement of apoE in AD is thus (1) of quantitative and not of
qualitative importance, because AD can develop without APOE4 and
might merely shift age of onset and (2) might depend upon environmental factors that are specific for industrialized countries.

2.2.1.4 ApoE Receptors

At least three apoE receptors, LDL receptor, LDL receptor-related protein (LRP) and glycoprotein 330 are known to be expressed in the brain,
and the former two have been shown to play a role in neurite growth
(Nathan et al. 1994; Kounnas et al 1995). ApoE4 displays normal
receptor binding to LDL receptor and LRP. In healthy brains, LRP is
found in neuronal cell bodies and the neuropil, rarely in astrocytes,
whereas in AD reactive astrocytes and βP deposits are stained additionally with antibodies against LRP. In contrast, almost no difference

between normal and AD brains was found for another apoE receptor, LDL receptor (Rebeck et al. 1993). Seven different and presumably unrelated LRP ligands (apoE, lactoferrin lipoprotein lipase, α2-macroglobulin, tissue and urokinase-type plasminogen activators, plasminogen activator inhibitor complexes) accumulate in βP deposits (Rebeck et al. 1995). In addition, N-terminal βAPP fragments (see below), in an isoform-dependent manner, are endocytosed and degraded via LRP (Kounnas et al. 1995). Glycoprotein 330 has at least some of the same ligands as LRP, most notably apoE, and apoJ, which is expressed at the apical surfaces of epithelia, suggesting a role in barrier function.

In concusion, LRP and possibly glycoprotein 330 may be involved in βA formation, maybe because trafficking of apoE or apoJ is disturbed.

2.2.2 Amyloid

"Amyloid," a term introduced by Virchow in 1854, is found extracellularly in form of aggregated fibrils with a β-pleated sheet conformation, the peptide backbone perpendicular to the fibril axis. Various proteins sharing certain histochemical and ultrastructural properties (e.g., green birefringence under polarized light after Congo-red binding) can become amyloid. The site of amyloid deposition in the body is strikingly restricted and probably depends upon the amyloid-forming protein itself rather than upon other factors equally found in those deposits. Here we will focus on amyloid deposited in the CNS.

Familial amyloidoses are associated with mutated amyloid-forming proteins, such as mutations of βP in hereditary cerebral hemorrhages with amyloidosis, Dutch type (HCHWA-D) and AD, cystatin C in HCHWA, Icelandic type (HCHWA-I), transthyretin in familial amyloid polyneuropathy, or prion protein in familial CJD and Gerstmann-Sträussler-Scheinker disease (GSS). In addition, sporadic amyloidoses without mutations of the amyloid-forming protein occur, such as in FAD and sporadic AD, sporadic and infectious prion diseases. βA and prion protein are predominantly deposited within the CNS and cerebrovasculature (Glenner and Wong 1984; Allsop et al. 1988), and cystatin C within the cerebrovasculature alone (Luyendijk et al. 1988); transthyretin is found in visceral organs and cerebrovasculature but not in brain parenchyma (Shirahama et al. 1981) and many other amyloid proteins are deposited systemically and rarely involve the CNS (e.g., Ishihara et al. 1989). In sporadic AD and FAD, a change in conformation of preamyloid noncongophilic β-peptide (βP) presumably gives rise to formation of congophilic βA fibrils.

2.2.2.1 βA Accumulation

βA constitutes the main protein component of vascular and parenchymal extracellular amyloid deposits found in all cases of AD, CAA, HCHWA-D and Down syndrome (DS). βA consists of an up to 43 amino acid long cleavage product of βAPP, the so-called βP, that is physiologically produced by many cells (Haass et al. 1992) and becomes protease resistant after aggregation into βA fibrils (Nordstedt et al. 1994; Knauer et al. 1992).

That βA formation plays an important role in AD is based on the following findings: (1) Cerebral βA deposits are found in all cases of AD; (2) βA but not βP is neurotoxic (Yankner et al. 1990) and activates microglia and astrocytes in vitro (Meda et al. 1995; Pike et al. 1994); (3) neurons surrounding βA are at greater risk of undergoing cell death, as assessed by detection of neuronal DNA fragmentation in AD autopsies (Lassmann et al. 1995); (4) AD with typical βA deposition together with increased βAPP expression due to an extra βAPP gene is found consistently in DS; and (5) different mutations on the βAPP gene are linked to FAD and HCHWA-D. These mutations are all within or adjacent to the βP sequence and result in cellular production of either more βP (βAPP670/671 mutation; Citron et al. 1992), higher percentages of longer and therefore more aggregable βP (βP42; βAPP717 mutation; Suzuki et al. 1994) or altered and therefore more aggregable βP (βAPP693 mutation within βP in HCHWA-D; Wisniewski et al. 1991). Cells with mutations of S182, another gene that is located on chromosome 14 and likely to cause most cases of early-onset FAD, produce more βP42 in vitro (Levy-Lahad et al. 1995). In chromosome 1-linked FAD, altered βP production has not yet been reported, but a high degree of homology with S182 suggests similar properties of the two gene products. Thus, so far all forms of FAD share one common property: Increased likelihood of βP aggregation into βA. Most AD patients suffer from sporadic AD where βA accumulation is always found but cannot be sufficiently explained yet. Possible mechanisms include increased production or decreased degradation of βP, increased rate of βP aggregation, or decreased degradation of βA (see below). The finding that degree of βP deposition does not correlate with degree of dementia and that cognitively normal elderly people show βP deposition quite frequently, although to a much lesser extent and rarely associated with neuritic changes, speaks against an important role of βA in AD. An explanation might be that βA is necessary but not sufficient for developing AD and that only when neurites are associated with βA, as in neuritic plaques, does dementia develop. This neuritic involvement is different from NFT, which might completely be lacking in AD (Terry et al. 1987).

2.2.2.2 βAPP Expression

The highly conserved βAPP gene is located on chromosome 21 and encodes several different isoforms from 365 to 770 amino acids (βAPP365 to βAPP770), derived from alternative splicing (Kang et al. 1987). βAPP mRNA is found in almost all human tissues (Tanzi et al. 1987) and is detected already during development (Fisher et al. 1991). In CNS, neurons, astrocytes, microglia, oligodendroglia, and vascular smooth muscle cells all produce βAPP: however, the predominant isoform varies (references in Sola et al. 1993). The βP sequence, located within the ectodomain–membrane transition of βAPP, is part of the four large isoforms βAPP 695, 714, 751, and 770 but not of βAPP-like proteins and βP can be generated by proteolytic cleavage of βAPP.

Since the transcription factor AP-1 is activated in response to various stimuli, including ischemia (Hsu et al. 1993; An et al. 1993), brain trauma (Dragunow et al. 1990), and axon damage (Jenkins and Hunt 1991), and the βAPP promoter contains an AP-1 binding site (Salbaum et al. 1988), βAPP expression may be upregulated in response to those stimuli. Indeed, many authors have found neuronal βAPP upregulation in animal models of global ischemia as assessed by immunostaining or in situ hybridization (Tomimoto et al. 1994, 1995; Heurteaux et al. 1993; Kalaria et al. 1993; Wakita et al. 1992). Where no neuronal βAPP upregulation is reported, the time of global ischemia has either been very long (≥20 min), with resultant neuronal death (Banati et al. 1995; Palacios et al. 1995; Robison et al. 1993), or the time of ischemia has been too short (2 min) and upregulation occurred only after prolonged (≥5 min) ischemic periods (Tomimoto et al. 1994). In agreement are the results in focal ischemia, where βAPP is found in peri-infarct neurons that probably suffered from an intermediate ischemic stress comparable to the global ischemia models (Stephenson et al. 1992). After 10 min of cardiac arrest, cerebral βAPP immunoreactivity is found intracellularly and extracellularly (Pluta et al. 1994). Human brains show βAPP immunoreactivity in infarct areas and βP immunoreactivity after head trauma, after hypoxic coma, and in arterial border zones (Nukina et al. 1992; Ohgami et al. 1992, Roberts et al 1991, Jendroska et al. 1996). Astrocytes are more resistant to ischemic damage and apparently slower in regard of βAPP induction. Three days after 5 min of global ischemia, βAPP accumulates in neurons but not in astrocytes (Tomimoto et al. 1995), whereas several days after global cerebral ischemia lasting 20 min or more, increased βAPP immunoreactivity is observed in hippocampal astrocytes while hippocampal neurons die (Banati et al. 1995; Palacios et al. 1995). In microglia activated after ischemic injury, increased βAPP immunoreactivity has been reported as well (Banati et al. 1995). Other types of injury with resultant βAPP upregulation include axotomy, kainate or ibotenate toxicity, cytokine or colchicine administration and heat shock (Sola et al. 1993, Shigematsu and McGeer 1992, Abe et al. 1991; Gray 1993). In contrast, metabolic stress due to physostigmine or glucocorticoid or diabetogenic treatment fails to elicit changes in βAPP expression (Wallace et al. 1993). Thus βAPP upregulation in different cell types is possible and depends upon degree and type of stimulus.

It has been speculated that permanent βAPP overexpression, as it is found in DS (Oyama et al. 1994), may be the reason for βA deposition in AD, but autopsy results are contradictory. Some authors describe an altered ratio of βAPP isoforms (Neve et al. 1990; Tanaka et al. 1988, 1989; Golde et al. 1990) or decrease in total βAPP mRNA (Clark et al. 1989; Goedert 1987), whereas others report no change in βAPP expression (König et al. 1991; Oyama et al. 1991). Johnson et al. (1990) found an association between βA deposition and βAPP expression while others did not (Golde et al. 1990; Hyman et al. 1993). However, because aggregation of βP into βA does not depend upon linear βP concentration but rather upon it reaching a critical level (Jarrett and Lansbury 1993), the occurrence of a single βP peak may be more critical for βA deposition than continuously moderately elevated βP concentration. Although it has been shown that the prevalence of βP deposits increases with age (Davies et al. 1988), it cannot be concluded that βP accumulates gradually over time. A single temporary "cerebral accident" (such as head trauma, ischemia, etc.) may be responsible for all βA found in an endstage AD brain and it may just be that the chance of experiencing such an event increases with age. Even if βP deposits do grow over a long time, seeding remains a critical event (Jarrett and Lansbury 1993). Therefore, because various cerebral injuries such as moderate cerebral ischemia or trauma cause a steep increase in βAPP expression, they possibly are a prime event in βA formation under the assumption that βP release increases concomitantly. In accordance with this concept is the observation that head trauma and cardiac disease have indeed been identified as risk factors in AD (Roberts et al. 1991; Aronson et al. 1990).

2.2.2.3 βAPP and βA Degradation

Proteolytic cleavage at various sites of βAPP can cause the release of variably sized ectodomain fragments of βAPP that have been shown to be neuroprotective (Mattson et al. 1993; Bowes et al. 1994), neurotrophic (Schubert et al. 1989), and essential for normal brain function (Müller et al. 1994). Brain injury is a strong stimulus for βAPP cleavage as assessed by release of ectodomain βAPP fragments during brain slice preparation (Farber et al. 1995). Release of βP depends upon βAPP cleavage and therefore might also be upregulated in this setting. Whether stimulated extracellular cleavage of βAPP within the βP domain reduces overall βP production of a cell in vitro depends upon the cell type used (Hung et al. 1993; Dyrks et al. 1994). βP production is inhibited by protein phosphorylation in cells transfected with βAPP (Buxbaum et al. 1993; Hung et al. 1993), whereas increased βP production is observed in cells from FAD patients carrying the βAPP670/671 mutation and after rise of intracellular calcium in cells transfected with βAPP (Querfurth and Selkoe 1994). Among CNS cells, astrocytes have been shown to produce highest levels of βP in vitro (Busciglio et al. 1993), whereas platelets are the major source in blood (Chen et al. 1995). Although βP is not harmful per se, it may contribute to βA formation, which appears to be important for AD pathogenesis in two ways: βP becomes neurotoxic and it becomes protease resistant. Aggregation is modulated by βP concentration and C-terminal length, point mutations and species-dependent sequence, interaction with preexisting βA, pH, oxidative environment, and presence or absence of cofactors such as apoE, transthyretin, apoJ, zinc, heparan sulfate and α1-antichy-motrypsin (Ma et al. 1994; Gallo et al. 1994; Schwarzman et al. 1994; Zlokovic et al. 1993; Fraser et al. 1992; Dyrks et al. 1992). It is likely that other factors are involved. Components found in βA deposits such as glycosaminoglycans or serum amyloid P component have been shown to further inhibit βA degradation in vitro (Tennent et al. 1995; Gupta-Bansal et al. 1995). In vivo, pure, synthetic βA can still be digested by phagocytes when injected into a rat brain, whereas βA injected together with proteoglycans can reside for many weeks (Fraut-schy et al. 1992; Snow et al. 1994).

In summary, increased or altered βAPP expression, altered βAPP processing, or altered βP trafficking are as likely as decreased βP degradation to contribute to AD pathogenesis by increasing the local concentration of βP and its likelihood to aggregate into βA. Aggregation kinetics predict that a single high dose of βP might be more important for amyloid formation than continuous moderate overproduction βA. βA, once formed, might then lead to neuronal challenge and glial activation.

2.2.2.4 Other Amyloid Proteins

A point mutation of cystatin C, a 110 amino acid lysosomal cysteine proteinase inhibitor, is linked to HCHWA-I (Palsdottir et al. 1988). Cystatin C is synthesized in the brain (as well in other organs) and has a very high CSF/plasma concentration ratio (Cole et al. 1989) that is decreased in HCHWA-I (Grubb and Lofberg 1985). Cystatin C-amyloid is found in the outer media or adventitia of arterioles and capillaries in leptomeninges and white and gray matter, but not in brain parenchyma. Clinically it resembles HCHWA-D with a much earlier age of onset in the second decade. Dementia is reported in 17/19 cases by Blöndal et al (1989). Parenchymal prion protein amyloid is found in in 100% of GSS disease, 75% of kuru, and 15% of CJD brains, all of which cause dementia (Roberts et al. 1988). Cerebrovascular prion protein (PP) amyloid in PP diseases has been reported in scrapie (Allsop et al. 1988). Two other diseases with cerebral congophilic deposits are familial amyloid polyneuropathy type I linked to a mutation of transthyretin, and cerebral autosomal dominant arteriopathy with subcortical infarcts and leukoencephalopathy (CADASIL) with deposition of a currently un-known substance. A linkage study has assigned the disease locus to chromosome 19q12 (Ragno et al. 1995). However, in both familial amyloid polyneuropathy type I and CADASIL, only cerebrovascular and not parenchymal congophilic deposits have been described and only CADASIL causes dementia.

Thus AD is the only cerebral amyloidosis with obligatory parenchymal amyloid deposits, suggesting that parenchymal amyloid may be necessary for disease progression. In HCHWA-I and prion diseases brain parenchymal amyloid may be absent while patients develop dementia, implying that at least in those diseases parenchymal amyloid deposition cannot be sufficient for development of the disease. Fibril formation without deposition in the extracellular matrix or vascular changes comparable to MID are possible alternative pathomechanisms.

Regional specificity of amyloid deposition in AD as compared to other amyloidoses is probably specific for βP rather than amyloid-associated compounds, because they are found within multiple forms of amyloidoses showing distinct patterns of amyloid deposition.

2.2.3 Interaction of ApoE and βA

In AD brains, apoE colocalizes with vascular and parenchymal βA deposits and even with noncongophilic (preamyloid) diffuse plaques, βP-negative meningeal vessels, and extracellular NFT (Namba et al. 1991; Yamaguchi et al. 1994). The extent of βP deposition but not the amount of apoE deposition correlates with the number of APOE4 alleles. After head trauma, increased βP deposition is found in the brains of APOE4 carriers but not in the brains of other APOE genotypes; however, a direct effect of apoE on βP synthesis could not be shown so far (Schmechel et al. 1993; Rebeck et al. 1993; Mayeux et al. 1995). After global ischemia, apoE and βAPP are upregulated simultaneously. Isoform-specific (apoE3>apoE4), oxidation-dependent, and uncommonly strong binding of apoE and βP/βA (see Fig. 1) has been demonstrated in vitro and in AD brain (Näslund et al. 1995; LaDu et al. 1995), providing an explanation for their colocalization. ApoE (apoE4>apoE3) accelerates aggregation of βP into βA (Wisniewski et al. 1994). As the apoE–βP complex does not easily cross the BBB, apoE in AD deposits may originate in the CNS rather than from an extracerebral source. Interestingly, both apoE and βP bind heparan sulfate (see Fig. 1), which is part of all AD amyloid deposits but lacks in noncongophilic βP deposits. Heparan sulfate, like apoE, not only binds to βP but also promotes βP aggregation and renders βA more resistant to degradation (Gupta-Bansal et al. 1995).

In summary, apoE4 may accelerate AD pathogenesis because of accelerated βA formation. Colocalization of apoE, heparan sulfate, and βP, e.g., at vascular basal membranes, possibly under low pH and free radical formation, provides a setting especially conducive to βP aggregation.

2.3 Cerebrovasculature in AD

2.3.1 Cerebral Amyloid Angiopathy

CAA is found in one half to one fourth of elderly individuals and is characterized by amyloid deposition in the cerebrovascular wall, which is, with rare exceptions (see below), βA. βA is deposited in the outer basal membrane perpendicular and without direct contact to the vessel wall. Progressing deposition spares the endothelium but may destroy the elastic lamina, media, adventitia, and smooth muscle cells and occlude the capillary lumen (Scholz 1938). Sites of deposition include small cerebral arteries and capillaries, whereas deposition of βA in large cerebral arteries, veins, small arteries, and capillaries outside the brain is sparse (Joachim et al. 1988). The great variety of diseases associated with CAA include (Vinters 1987):

- AD
- Sporadic hemorrhage
- HCHWA-D/I
- GSS disease
- Dementia pugilistica
- Cereballar ataxia
- Cerebral vasculitis
- Postradiation necrosis of the brain
- DS
- Cerebral microinfarcts
- Cerebrovascular malformation

In contrast to most cases of AD, CAA patients develop focal neurological deficits from the beginning, most likely due to cortical and white matter hemorrhages that cause cerebral infarctions, transient ischemic attacks (TIAs), or focal seizures. CAA accounts for 5%–10% of primary nontraumatic brain hemorrhages, which account for 10% of all strokes. Hemorrhages usually occur in the cortex and subcortical white matter, preferentially in frontal lobes (see Vinters 1987 for review). In patients without major hemorrhage, transient neurological symptoms or, in 30% of cases, rapidly progressing dementia may develop, correlating with the number of cortical hemorrhages. Parenchymal βA deposits and NFT are sparse. Presumed triggers of CAA include head injury, heart failure, surgery, and hypoxia. Risk factors for CAA are apoE4, AD, advancing age, inheritance of an APP-mutation (HCHWA-D), CC mutation (HCHWA-I), or a transthyretin mutation (type I familial polyneuropathy). CAA is not associated with hypertension, diabetes, or atherosclerosis. Several diseases are associated with CAA (see list of diseases above).

HCHWA-D, a rare familial disorder that is caused by a mutation of βAPP within the βP-sequence, is of particular interest for AD. Patients show congophilic vascular βA deposition but only diffuse, noncongophilic parenchymal βA without there being an underlying neuritic condition (Timmers et al. 1990). The first hemorrhage is lethal in two out of three cases, with a mean age of onset of 52 years (Luyendijk et al. 1988). Dementia has been reported even without previous strokes (Haan et al. 1992; van Duinen et al. 1987), indicating that dementia may result from CAA by mechanisms independent of hemorrhage.

The great variety of triggers and diseases associated with CAA might offer clues to the genesis of parenchymal βA deposits that are, after all, made up of the same protein. Multiple infarcts may cause dementia in HCHWA-D; however, dementia in HCHWA-D is possible in the absence of strokes and congophilic parenchymal amyloid while CAA without hemorrhages occurs in most cases of dementia of AD type, which may imply that vascular changes not yet understood play a role in βA-related dementias.

2.3.2 Vasculature and βA in AD

By definition all AD brains show parenchymal βA but, in addition, most, if not all, patients with AD show focal vascular βA deposits in the form of CAA. Because the main protein components of vascular and parenchymal amyloid share an identical sequence, a causal relationship between the two has been sought for. Lippa et al (1993) found that sites of maximal vascular and parenchymal βA load in AD brains are different, while Rosenblum and Haider (1988) found a negative correlation between the amount of βA deposited. Whether βA has a vascular or parenchymal source is another point of interest. Miyakawa et al (1992) found at least one capillary within a plaque, whereas Kawai et al (1992) found only 8%–13% of plaques penetrated by a capillary, the highest capillary density in its border. Isomerization and racemization, protein modifications that increase over time, were found on parenchymal rather than on vascular βA by Roher et al (1993), which might indicate that parenchymal βA precedes vascular βA. This notion is supported by cases of DS in which parenchymal βA without concomitant vascular βA is found (Iwatsubo et al. 1995).

Platelets containing βAPP in α-granules are the major source of βP in blood and might therefore contribute to βA deposition if βP crosses the BBB (Bush et al. 1990; Chen et al. 1995). βP was found to promote platelet aggregation, decrease platelet membrane fluidity, and stimulate tissue-type plasminogen activator, a thrombolytic agent that is normally stimulated by fibrin (Kowalska and Badellino 1994; Müller et al. 1995; Kingston et al. 1995) and is found within βA deposits.

In conclusion, the relationship between vascular and parenchymal deposits remains to be elucidated, and vascular contributions in the form of βP leakage through the BBB cannot be excluded as being important for parenchymal changes in AD.

2.3.3 BBB and Brain Capillaries

The BBB serves as barrier and carrier and is located at the capillary endothelium. The barrier function is due to tight junctions and diminished transcytosis of endothelial cells supported by pinocytes, astrocytes, and neurons (Pardridge 1988; Mooradian 1988). In AD, a variety of capillary changes have been reported, including CAA (see Sect. 2.3.1), surface abnormalities (Scheibel et al. 1989), loss of perivascular nerve plexus innervated by neurons from locus caeruleus, nucleus basalis, and dorsalis raphe, upregulation of adrenergic receptors, and reduction of glucose carrier (Mann et al. 1992; Kalaria and Harik 1989). In general, barrier disruption results in leakage of large molecules from the plasma into the brain and is assessed by dye uptake, increased CSF/serum ratio of serum proteins, or detection of serum proteins in the brain (e.g., albumin). In most studies, dye leakage or increased CSF/serum ratio cannot be found in AD, in contrast to VDs, and only certain plasma proteins such as immunoglobulins or complement factors (but not albumin) are detected in AD brains (for references see Stewart et al. 1992; Rozemuller et al. 1988).

βP is able to cross the BBB, a process facilitated by apoJ binding and inhibited by apoE binding or glycoprotein 330 antibodies (Zlokovic et al. 1993, 1994, 1995). ApoJ has structural resemblance to apoE and is predominantly expressed at fluid–tissue interfaces. In plasma, apoJ–βP complexes are found, preferentially in HDL3 and very low density lipoprotein (Koudinov et al. 1994; see May and Finch 1992 for review). In the CNS apoJ is detected in epithelial cells of the plexus choroideus and ependymal cells, but also in a subset of neurons and glial cells, where it can be upregulated in response to experimental entorhinal lesions, cerebral kainate injection, and global ischemia. ApoJ interacts with immunoglobulin and complement factors, both serum proteins found in AD brain deposits, but not in normal brains, and is the major βP-binding protein in human CSF (Aronow et al. 1993). ApoJ, which is upregulated and found in βP deposits, likely plays a role in AD. For cell death apoJ is presumably more protective than destructive and might fail to fulfill this task in AD, possibly due to lowered affinity of apoJ for βP after βP aggregation (Matsubara et al. 1995).

In summary, a major leakage of BBB in AD does not seem very likely, but other changes in capillary properties might still play a role, for example, through narrowing of the capillary lumen and loss of tone regulation with subsequent malnutrition of the brain or altered βP transport mediated by apoJ and glycoprotein 330.

2.4 Imaging Findings and Morphological Correlates

Although a wealth of imaging studies on gray matter changes in AD have been conducted, we decided to concentrate on white matter changes in AD because gray matter changes are more likely due to primary neuronal loss. Vascular changes in these regions, detected as decreased cerebral blood flood or glucose or oxygen consumption, seem to be more a consequence than a cause of dementia in AD. In contrast, white matter changes on computed tomography or magnetic resonance imaging scans, which are found to be more severe in AD than in age-matched controls in most studies, may contribute to AD. Excess white matter changes in AD patients before onset of dementia and in healthy offspring of AD patients supports this hypothesis (Bowen et al. 1990; Coffman et al. 1990).

Many different histologic alterations may be the basis for white matter changes, including arteriolosclerosis, dilated perivascular spaces, loss of axons and myelin, gliosis, and lacunar infarction (see references in Scheltens et al. 1995). Because the arterioles in cerebral white matter are not arranged in collaterals and thus irreplaceable in case of dysfunction, loss of surrounding axons results (Hachinski et al. 1987). This indeed correlates with white matter changes in AD (Scheltens et al. 1995; Waldemar et al. 1994; Leys et al. 1991). Three pathomechanisms could be responsible: wallerian degeneration due to cortical atrophy ((Leys et al. 1991), incomplete infarction due to hypoperfusion (Brun and Englund 1986), or incomplete infarction due to CAA. Hypoperfusion may be a primary event but also secondary to CAA, compromising small vessels. CAA is more pronounced in gray matter, but is also detected within white matter (Tabaton et al. 1991; Vinters et al. 1990; Kawai et al. 1990). In addition, white matter changes colocalize with βA deposits (Janota et al. 1989; Vinters et al. 1991), both correlate in extent (Tabaton et al. 1991), and white matter changes in AD and hemorrhagic CAA histologically resemble incomplete infarctions (Brun and Englund 1986; Haan et al. 1990; for exceptions see Rezek et al. 1987), making CAA a likely explanation for white matter changes in AD. In hemorrhagic CAA, where white matter changes occur without there being an underlying neuritic condition (Haan et al. 1991), wallerian degeneration seems unlikely, but is a possible cause of white matter changes in AD.

The significance of white matter changes in the pathogenesis of AD has not been clearly established, but various, though quite different correlations, have been found, including age of onset, cognitive performance, decreased motoric efficiency, increase in gait disturbances, or earlier death (for review see Duara 1994).

White matter changes, possibly due to CAA, are certainly not sufficient to cause full-blown AD, but are clearly more severe in AD than in normal aging. If detected very early, they correlate with clinical symptoms and might therefore add a VD component to the dementia of parenchymal origin predominating in AD.

2.5 Conclusion

There is a fine line between neuronal and vascular pathology in AD and related disorders, as exemplified by the involvement of two genes in both processes. Dichotomy of βAPP culminates in a Dutch pedigree, where either AD or CAA develops in family members who carry the same point mutation. The second dichotomy is observed with apoE. APOE4 frequencies are increased in vascular and neuronal disease, and in AD apoE4 may act as an accelerator of disease progress, depending upon environmental factors. The formation of βA may be a critical event in the pathogenesis of AD. The fact that this process is accelerated by apoE might link the two diseases.

Because sporadic AD develops in the absence of βAPP mutations and is not dependent upon APOE4, additional factors ought to play a role. Head trauma has been shown to be a risk factor in AD and for the following reasons cerebral ischemia might be another one: Cerebral ischemia (1) causes upregulation of βAPP, (2) induces various factors potentially involved in βP aggregation (apoE, acute phase proteins, low pH, free radicals), and (3) might cause proteolytic cleavage of βAPP into amyloidogenic fragments. Thus cerebral ischemia, possibly occurring without immediate consequences or symptoms, may start a process that becomes symptomatic years or decades later in the form of AD by leaving the same remnants in form of βP deposits behind. Microvascular changes found in AD brains, possibly induced by vascular βA deposition, may contribute to AD pathogenesis by causing ischemic conditions.

References

Abe K, Tanzi RE, Kogure K (1991) Selective induction of Kunitz-type protease inhibitor domain-containing amyloid precursor protein mRNA after persistent focal ischemia in rat cerebral cortex. Neurosci Lett 125(2):172–174

Allsop D, Ikeda S, Bruce M, Glenner GG (1988) Cerebrovascular amyloid in scrapie-affected sheep reacts with antibodies to prion protein. Neurosci Lett 92:234–239

Alzheimer's Disease Collaborative Group (1993) Apolipoprotein E genotype and Alzheimer's disease. Lancet 342:737–738 (letter)

Amouyel P, Vidal O, Launay JM, Laplanche JL (1994) The apolipoprotein E alleles as major susceptibility factors for Creutzfeldt-Jakob disease. The French Research Group on Epidemiology of Human Spongiform Encephalopathies. Lancet 344:1315–1318

An G, Lin TN, Liu JS, Xue JJ, He YY, Hsu CY (1993) Expression of c-fos and c-jun family genes after focal cerebral ischemia. Ann Neurol 33:457–464

Aronow BJ, Lund SD, Brown TL, Harmony JA, Witte DP (1993) Apolipoprotein J expression at fluid-tissue interfaces: potential role in barrier cytoprotection. Proc Natl Acad Sci USA 90:725–729

Aronson MK, Ooi WL, Morgenstern H, Hafner A, Masur D, Crystal H, Fisher D, Katzman R (1990) Women, myocardial infarction, and dementia in the very old. Neurology 40:1102–1106

Banati RB, Gehrmann J, Wiessner C, Hossmann KA, Kreutzberg GW (1995) Glial expression of the beta-amyloid precursor protein (APP) in global ischemia. J Cereb Blood Flow Metab 15:647–654

Blöndal H, Guomundsson G, Benedikz E, Johannesson G (1989) Dementia in hereditary cystatin C amyloidosis. Prog Clin Biol Res 317:157–164

Bowen BC, Barker WW, Loewenstein DA, Sheldon J, Duara R (1990) MR signal abnormalities in memory disorder and dementia. AJR Am J Roentgenol 154:1285–1292

Bowes MP, Masliah E, Otero DA, Zivin JA, Saitoh T (1994) Reduction of neurological damage by a peptide segment of the amyloidbeta/A4 protein precursor in a rabbit spinal cord ischemia model. Exp Neurol 129:112–119

Boyles JK, Pitas RE, Wilson E, Mahley RW, Taylor JM (1985) Apolipoprotein E associated with astrocytic glia of the central nervous system and with nonmyelinating glia of the peripheral nervous system. J Clin Invest 76(4):1501–1513

Boyles JK, Zoellner CD, Anderson LJ, Kosik LM, Pitas RE, Weisgraber KH, Mahley RW, Gebicke Haerter PJ, Ignatius MJ, Shooter EM (1989) A role for apolipoprotein E, apolipoprotein A-I, and low density lipoprotein receptors in cholesterol transport during regeneration and remyelination of the rat sciatic nerve. J Clin Invest 83:1015–1031

Boyles JK, Notterpek LM, Anderson LJ (1990) Accumulation of apolipoproteins in the regenerating and remyelinating mammalian peripheral nerve. Identification of apolipoprotein D, apolipoprotein-IV, apolipoprotein E, and apolipoprotein A-I. J Biol Chem 265(29):17805–17815

Brun A, Englund E (1986) A white matter disorder in dementia of the Alzheimer type: a pathoanatomical study. Ann Neurol 19:253–262

Busciglio J, Gabuzda DH, Matsudaira P, Yankner BA (1993) Generation of beta-amyloid in the secretory pathway in neuronal and nonneuronal cells. Proc Natl Acad Sci USA 90:2092–2096

Bush AI, Martins RN, Rumble B, Moir R, Fuller S, Milward E, Currie J, Weidemann A, Fischer P, Multhaup G, Beyreuther K, Masters CL (1990) The amyloid precursor protein of Alzheimer's disease is released by human platelets. J Biol Chem 265:15977–15983

Buxbaum JD, Koo EH, Greengard P (1993) Protein phosphorylation inhibits production of Alzheimer amyloid beta/A4peptide. Proc Natl Acad Sci USA 90:9195–9198

Carlsson J, Armstrong VW, Reiber H, Felgenhauer K, Seidel D (1991) Clinical relevance of the quantification of apolipoprotein E in cerebrospinal fluid. Clin Chim Acta 196:167–176

Chen M, Inestrosa NC, Ross GS, Fernandez HL (1995) Platelets are the primary source of amyloid beta-peptide in human blood. Biochem Biophys Res Commun 213(1):96–103

Citron M, Oltersdorf T, Haass C, McConlogue L, Hung AY, Seubert P, Lieberburg I, Selkoe DJ (1992) Mutation of the beta-amyloid precursor protein in familial Alzheimer's disease increases beta-protein production. Nature 360:672–674

Clark AW, Krekoski CA, Parhad IM, Liston D, Julien JP, Hoar DI (1989) Altered expression of genes for amyloid and cytoskeletal proteins in Alzheimer cortex. Ann Neurol 25:331–339

Coffman JA, Torello MW, Bornstein RA, Chakeres D, Burns E, Nasrallah HA (1990) Leukoaraiosis in asymptomatic adult offspring of individuals with Alzheimer's disease. Biol Psychiatry 27:1244–1248

Cole T, Dickson PW, Esnard F, Averill S, Risbridger GP, Gauthier F (1989) The cDNA structure and expression analysis of the genes for the cysteine proteinase inhibitor cystatin C and for beta 2-microglobulin in rat brain. Eur J Biochem 186:35–42

Davies L, Wolska B, Hilbich C, Multhaup G, Martins R, Simms G, Beyreuther K, Masters Cl (1988) A4 amyloid protein deposition and the diagnosis of Alzheimer's disease: prevalence in aged brains determined by immunocytochemistry compared with conventional neuropathologic techniques. Neurology 38(11):1688–1693

Davignon J, Gregg RE, Sing CF (1988) Apolipoprotein E polymorphism and atherosclerosis. Arteriosclerosis 8:1–21

Dragunow M, Goulding M, Faull RL, Ralph R, Mee E, Frith R (1990) Induction of c-fos mRNA and protein in neurons and glia after traumatic brain injury: pharmacological characterization. Exp Neurol 107:236–248

Duara R (1994) Neuroimaging with CT and MRI in Alzheimer Disease. In: Terry RD, Katzman R, Bick KL (eds) Alzheimer disease. Raven, New York, pp 75–86

Dyrks T, Dyrks E, Hartmann T, Masters C, Beyreuther K (1992) Amyloidogenicity of beta A4 and beta A4-bearing amyloid protein precursor fragments by metal-catalyzed oxidation. J Biol Chem 267(25):18210–18217

Dyrks T, Monning U, Beyreuther K, Turner J (1994) Amyloid precursor protein secretion and beta A4 amyloid generation are not mutually exclusive. FEBS Lett 349:210–214

Farber SA, Nitsch RM, Schulz JG, Wurtman R. (1995) Regulated secretion of beta-amyloid precursor protein in rat brain. J Neurosci 15(11):7442–7451

Fisher S, Gearhart JD, Oster Granite ML (1991) Expression of the amyloid precursor protein gene in mouse oocytes and embryos. Proc Natl Acad Sci USA 88:1779–1782

Fraser PE, Nguyen JT, Chin DT, Kirschner DA (1992) Effects of sulfate ions on Alzheimer beta/A4 peptide assemblies: implications for amyloid fibril-proteoglycan interactions. J Neurochem 59:1531–1540

Frautschy SA, Cole GM, Baird A (1992) Phagocytosis and deposition of vascular beta-amyloid in rat brains injected with Alzheimer beta-amyloid. Am J Pathol 140:1389–1399

Frisoni GB, Bianchetti A, Govoni S, Trabucchi M, Calabresi L (1994) Association of apolipoprotein E E4 with vascular dementia. JAMA 271:1317

Gallo G, Wisniewski T, Choi Miura NH, Ghiso J, Frangione B (1994) Potential role of apolipoprotein-E in fibrillogenesis. Am J Pathol 145:526–530

Ghiselli G, Schaefer EJ, Zech LA, Gregg RE (1982) Increased prevalence of apolipoprotein E4 in type V hyperlipoproteinemia. J Clin Invest 70:474–477

Glenner GG, Wong CW (1984) Alzheimer's disease: initial report of the pufication and characterization of a novel cerebrovascular amyloid protein. Biochem Biophys Res Commun 120:885–890

Goedert M (1987) Neuronal localization of amyloid beta protein precursor mRNA in normal human brain and in Alzheimer's disease. EMBO J 6:3627–3632

Golde TE, Estus S, Usiak M, Younkin LH, Younkin SG (1990) Expression of beta amyloid protein precursor mRNAs: recognition of a novel alternatively spliced form and quantitation in Alzheimer's disease using PCR. Neuron 4:253–267

Gray CW (1993) Regulation of beta-amyloid precursor protein isoform mRNAs by transforming growth factor-beta 1 and interleukin-1 beta in astrocytes. Brain Res Mol Brain Res 19:251–256

Greenberg SM, Rebeck GW, Vonsattel JPG, Gomez-Isla T, Hyman BT (1995) Apolipoprotein E4 and cerebral hemorrhage associated with amyloid angiopathy. Ann Neurol 38(2):254–259

Grubb A, Lofberg H (1985) Human gamma-trace. Structure, function and clinical use of concentration measurements. Scand J Clin Lab Invest Suppl 177:7–13

Gupta-Bansal R, Frederickson RA, Brunden K (1995) Proteoglycan mediated inhibition of A-beta proteolysis: a potential cause of senile plaque accumulation. J Biol Chem 270(31):18666–18671

Haan J, Roos RA, Algra PR, Lanser JB, Bots GT, Vegter Van der Vlis M (1990) Hereditary cerebral haemorrhage with amyloidosis – Dutch type. Magnetic resonance imaging findings in 7 cases. Brain 113:1251–1267

Haan J, Hardy JA, Roos RA (1991) Hereditary cerebral hemorrhage with amyloidosis – Dutch type: its importance for Alzheimer research. Trends Neurosci 14:231–234

Haan J, Bakker E, Jennekens Schinkel A, Roos RA (1992) Progressive dementia, without cerebral hemorrhage, in a patient with hereditary cerebral amyloid angiopathy. Clin Neurol Neurosurg 94:317–318

Haan J, Maat Schieman ML, van Duinen SG, Jensson O, Thorsteinsson L (1994) Co-localization of beta/A4 and cystatin C in cortical blood vessels in Dutch, but not in Icelandic hereditary cerebral hemorrhage with amyloidosis. Acta Neurol Scand 89:367–371

Haass C, Schlossmacher MG, Hung AY, Vigo Pelfrey C, Mellon A, Lieberburg I, Koo EH, Schenk D, Teplow DB, Selkoe DJ (1992) Amyloid beta-peptide is produced by cultured cells during normal metabolism. Nature 359:322–325

Hachinski VC, Potter P, Merskey H (1987) Leuko-araiosis. Arch Neurol 44:21–23

Hall ED, Oostveen JA, Dunn E, Carter DB (1995) Role of amyloid protein precursor and apolipoprotein E in stroke. In: Moskowitz MA, Caplan LR (eds) Cerebrovascular diseases. Nineteenth Princeton Stroke Conference. Butterworth-Heinemann, Boston, pp 251–262

Han SH, Einstein G, Weisgraber KH, Strittmatter WJ, Saunders AM, Roses AD, Schmechel DE (1994) Apolipoprotein E is localized to the cytoplasm of human cortical neurons: a light and electron microscopic study. J Neuropathol Exp Neurol 53:535–544

Heurteaux C, Bertaina V, Widmann C, Lazdunski M (1993) K+ channel openers prevent global ischemia-induced expression of c-fos,c-jun, heat shock protein, amyloid beta-protein precursor genes and neuronal death in rat hippocampus. Proc Natl Acad Sci USA 90:9431–9435

Hsu CY, An G, Liu JS, Xue JJ, He YY, Lin TN (1993) Expression of immediate early gene and growth factor mRNAs in a focal cerebral ischemia model in the rat. Stroke 24:178

Hung AY, Haass C, Nitsch RM, Qiu WQ, Citron M, Wurtman RJ, Growdon JH (1993) Activation of protein kinase C inhibits cellular production of the amyloid beta-protein. J Biol Chem 268:22959–22962

Hyman BT, Wenniger JJ, Tanzi RE (1993) Nonisotopic in situ hybridization of amyloid beta protein precursor in Alzheimer's disease: expression in neurofibrillary tangle bearing neurons and in the microenvironment surrounding senile plaques. Brain Res Mol Brain Res 18:253–258

Ishihara T, Nagasawa T, Yokota T, Gondo T, Takahashi M, Uchino F (1989) Amyloid protein of vessels in leptomeninges, cortices, choroid plexuses, and pituitary glands from patients with systemic amyloidosis. Hum Pathol 20:891–895

Iwatsubo T, Mann DM, Odaka A, Suzuki N, Ihara Y (1995) Amyloid beta protein (A beta) deposition: A beta 42(43) precedes A beta 40 in Down syndrome. Ann Neurol 37:294–299

Janota I, Mirsen TR, Hachinski VC, Lee DH, Merskey H (1989) Neuropathologic correlates of leuko-araiosis. Arch Neurol 46:1124–1128

Jarrett JT, Lansbury PT (1993) Seeding "one-dimensional crystallization" of amyloid: a pathogenic mechanism in Alzheimer's disease and scrapie? Cell 73:1055–1058

Jendroska K, Poewe W, Daniel SE, Pluess J (1996) Ischemic stress induces deposition of amyloid beta immunoreactivity in human brain. Acta Neuropathol (in press)

Jenkins R, Hunt SP (1991) Long-term increase in the levels of c-jun mRNA and jun protein-like immunoreactivity in motor and sensory neurons following axon damage. Neurosci Lett 129:107–110

Joachim CL, Morris JH, Selkoe DJ (1988) Clinically diagnosed Alzheimer's disease: autopsy results in 150 cases. Ann Neurol 24:50–56

Johnson SA, McNeill T, Cordell B, Finch CE (1990) Relation of neuronal APP-751/APP-695 mRNA ratio and neuritic plaque density in Alzheimer's disease. Science 248:854–857

Kalaria RN, Harik SI (1989) Abnormalities of the glucose transporter at the blood-brain barrier and in brain in Alzheimer's disease. Prog Clin Biol Res 317:415–421

Kalaria RN, Bhatti SU, Lust WD, Perry G (1993) The amyloid precursor protein in ischemic brain injury and chronic hypoperfusion. Ann NY Acad Sci 695:190–193

Kang J, Lemaire HG, Unterbeck A, Salbaum JM, Masters CL, Grzeschik KH, Beyreuther K, Müller Hill B (1987) The precursor of Alzheimer's disease amyloid A4 protein resembles a cell-surface receptor. Nature 325:733–736

Kawai M, Kalaria RN, Harik SI, Perry G (1990) The relationship of amyloid plaques to cerebral capillaries in Alzheimer's disease. Am J Pathol 137(6):1435–1446

Kawai M, Cras P, Perry G (1992) Serial reconstruction of beta-protein amyloid plaques: relationship to microvessels and size distribution. Brain Res 592:278–282

Kingston IB, Castro MJM, Anderson S (1995) In vitro stimulation of tissue-type plasminogen activator by Alzheimer amyloid beta-peptide analogues. Nature Med 1:138–142

Knauer MF, Soreghan B, Burdick D, Kosmoski J, Glabe CG (1992) Intracellular accumulation and resistance to degradation of the Alzheimer amyloid A4/beta protein. Proc Natl Acad Sci USA 89:7437–7441

König G, Salbaum JM, Wiestler O, Lang W, Schmitt HP, Masters CL (1991) Alternative splicing of the beta A4 amyloid gene of Alzheimer's disease in cortex of control and Alzheimer's disease patients. Brain Res Mol Brain Res 9:259–262

Koudinov A, Matsubara E, Frangione B, Ghiso J (1994) The soluble form of Alzheimer's amyloid beta protein is complexed to high density lipoprotein 3 and very high density lipoprotein in normal human plasma. Biochem Biophys Res Commun 205:1164–1171

Kounnas MZ, Moir RD, Rebeck GW, Bush AI, Argraves WS, Tanzi RE, Hyman BT (1995) LDL receptor-related protein, a multifunctional ApoE receptor, binds secreted beta-amyloid precursor protein and mediates its degradation. Cell 82:331–340

Kowalska MA, Badellino K (1994) Beta-amyloid protein induces platelet aggregation and supports platelet adhesion. Biochem Biophys Res Commun 205:1829–1835

LaDu MJ, Pederson TM, FrailDE, Reardon CA, Getz GS, Falduto MT (1995) Purification of apolipoprotein E attentuates isoform-specific binding to beta-amyloid. J Biol Chem 270(16):9039–9042

Lassmann H, Bancher C, Breitschopf H, Wegiel J, Bobinski M, Jellinger K (1995) Cell death in Alzheimer's disease evaluated by DNA fragmentation in situ. Acta Neuropathol (Berl) 89:35–41

Levy-Lahad E, Wasco W, Poorkaj P, Romano DM, Oshima J, Pettingell WH, Jondro PD, Schmidt SD, Wang K, Crowley AC, Fu Y, Guenette SY, Galas D, Nemens E, Wijsman EM, Bird TD, Schellenberg GD and Tanzi RE (1995) Candidate gene for the chromosome 1 familial Alzheimer's disease locus. Science 269:973–977

Leys D, Pruvo JP, Parent M, Vermersch P, Soetaert G, Steinling M, Defossez A, Rapoport A, Clarisse J, Petit H (1991) Could Wallerian degeneration contribute to "leuko-araiosis" in subjects free of any vascular disorder? J Neurol Neurosurg Psychiatry 54:46–50

Lippa CF, Hamos JE, Smith TW, Pulaski Salo D, Drachman DA (1993) Vascular amyloid deposition in Alzheimer's disease. Neither necessary nor sufficient for the local formation of plaques or tangles. Arch Neurol 50:1088–1092

Luyendijk W, Bots GT, Vegter Van der Vlis M, Went LN, Frangione B (1988) Hereditary cerebral haemorrhage caused by cortical amyloid angiopathy. J Neurol Sci 85:267–280

Ma J, Yee A, Brewer HB, Jr., Das S, Potter H (1994) Amyloid-associated proteins alpha 1-antichymotrypsin and apolipoprotein E promote assembly of Alzheimer beta-protein into filaments. Nature 372:92–94

Mahley RW (1988) Apolipoprotein E: cholesterol transport protein with expanding role in cell biology. Science 240:622–630

Majack RA, Castle CK, Goodman LV, Weisgraber KH, Mahley RW, Shooter EM, Gebicke Haerter PJ (1988) Expression of apolipoprotein E by cultured vascular smooth muscle cells is controlled by growth state. J Cell Biol 107(3):1207–1213

Mann DM, Purkiss MS, Bonshek RE, Jones D, Brown AM, Stoddart RW (1992) Lectin histochemistry of cerebral microvessels in ageing, Alzheimer's disease and Down's syndrome. Neurobiol Aging 13:137–143

Mattson MP, Cheng B, Culwell AR, Esch FS, Lieberburg I, Rydel RE (1993) Evidence for excitoprotective and intraneuronal calcium-regulating roles for secreted forms of the beta-amyloid precursor protein. Neuron 10:243–254

Matsubara E, Frangione B, Ghiso J (1995) Characterization of apolipoprotein J-Alzheimer's A beta interaction. J Biol Chem 270(13):7563–7567

May PC, Finch CE (1992) Sulfated glycoprotein 2: new relationships of this multifunctinal protein to neurodegeneration. Trends Neurosci 15(10):391–396

Mayeux R, Ottman R, Maestre G, Ngai C, Tang M, Ginsberg H, Chun M, Tycko B, Shelanski M (1995) Synergistic effects of traumatic head injury and apolipoprotein-epsilon 4 in patients with Alzheimer's disease. Neurology 45:555–557

Meda L, Cassatella MA, Szendrei GI, Otvos L, Jr., Baron P, Villalba M, Rossi F (1995) Activation of microglial cells by beta-amyloid protein and interferon-gamma. Nature 374:647–650

Miyakawa T, Katsuragi S, Yamashita K, Ohuchi K (1992) Morphological study of amyloid fibrils and preamyloid deposits in the brain with Alzheimer's disease. Acta Neuropathol (Berl) 83:340–346

Mooradian AD (1988) Effect of aging on the blood-brain barrier. Neurobiol Aging 9:31–39

Müller U, Cristina N, Li ZW, Wolfer DP, Lipp HP, Rulicke T, Brandner S, Weissmann C (1994) Behavioral and anatomical deficits in mice homozygous for a modified beta-amyloid precursor protein gene. Cell 79:755–765

Müller WE, Koch S, Eckert A, Hartmann H, Scheuer K (1995) Beta-amyloid peptide decreases membrane fluidity. Brain Res 674:133–136

Namba Y, Tomonaga M, Kawasaki H, Otomo E, Ikeda K (1991) Apolipoprotein E immunoreactivity in cerebral amyloid deposits and neurofibrillary tangles in Alzheimer's disease and kuru plaque amyloid in Creutzfeldt-Jakob disease. Brain Res 541:163–166

Näslund J, Thyberg J, Tjernberg LO, Wernstedt C, Karlstrom AR, Gandy SE, Lannfelt L, Terenius L, Nordstedt C (1995) Characterization of stable complexes involving apolipoprotein E and the amyloid beta peptide in Alzheimer's disease brain. Neuron 15:219–228

Nathan BP, Bellosta S, Sanan DA, Weisgraber KH, Mahley RW, Pitas RE (1994) Differential effects of apolipoproteins E3 and E4 on neuronal growth in vitro. Science 264:850–852

Neve RL, Rogers J, Higgins GA (1990) The Alzheimer amyloid precursor-related transcript lacking the beta/A4sequence is specifically increased in Alzheimer's disease brain. Neuron 5:329–338

Noguchi S, Murakami K, Yamada N (1993) Apolipoprotein E genotype and Alzheimer's disease. Lancet 342:737

Nordstedt C, Naslund J, Tjernberg LO, Karlstrom AR, Thyberg J, Terenius L (1994) The Alzheimer A beta peptide develops protease resistance in association with its polymerization into fibrils. J Biol Chem 269:30773–30776

Nukina N, Kanazawa I, Mannen T, Uchida Y (1992) Accumulation of amyloid precursor protein and beta-protein immunoreactivities in axons injured by cerebral infarct. Gerontology 38 [Suppl 1]:10–14

Ohgami T, Kitamoto T, Tateishi J (1992) Alzheimer's amyloid precursor protein accumulates within axonal swellings in human brain lesions. Neurosci Lett 136:75–78

Osuntokun BO, Sahota A, Ogunniyi AO, Gureje O, Baiyewu O, Adeyinka A, Oluwole SO, Komolafe O, Hall KS, Unverzagt FW, Hui SL, Yang M, Hendrie HC (1995) Lack of an association between apolipoprotein e epsilon4 and Alzheimer's disease in elderly Nigerians. Ann Neurol 38:463–465

Oyama F, Shimada H, Oyama R, Titani K, Ihara Y (1991) Differential expression of beta amyloid protein precursor (APP) and tau mRNA in the aged human brain: individual variability and correlation. J Neuropathol Exp Neurol 50:560–578

Oyama F, Cairns NJ, Shimada H, Oyama R, Titani K, Ihara Y (1994) Down's syndrome: up-regulation of beta-amyloid protein precursor and tau mRNAs and their defective coordination. J Neurochem 62:1062–1066

Palacios G, Mengod G, Tortosa A, Ferrer I, Palacios JM (1995) Increased beta-amyloid precursor protein expression in astrocytes in the gerbil hippocampus following ischaemia: association with proliferation of astrocytes. Eur J Neurosci 7:501–510

Palsdottir A, Abrahamson M, Thorsteinsson L, Arnason A, Olafsson I, Jensson O (1988) Mutation in cystatin C gene causes hereditary brain haemorrhage. Lancet 2:603–604

Pardridge WM (1988) Does the brain's gatekeeper falter in aging? Neurobiol Aging 9:44–46

Pike CJ, Cummings BJ, Monzavi R, Cotman CW (1994) Beta-amyloid-induced changes in cultured astrocytes parallel reactive astrocytosis associated with senile plaques in Alzheimer's disease. Neuroscience 63:517–531

Plump AS, Smith JD, Hayek T, Aalto Setala K, Walsh A, Verstuyft JG, Breslow JL (1992) Severe hypercholesterolemia and atherosclerosis in apolipoprotein E-deficient mice created by homologous recombination in ES cells. Cell 71:343–353

Pluta R, Kida E, Lossinsky AS, Golabek AA, Mossakowski MJ, Wisniewski HM (1994) Complete cerebral ischemia with short-term survival in rats induced by cardiac arrest. I. Extracellular accumulation of Alzheimer's beta-amyloid protein precursor in the brain. Brain Res 649:323–328

Popko B, Goodrum JF, Bouldin TW, Zhang SH, Maeda N (1993) Nerve regeneration occurs in the absence of apolipoprotein E in mice. J Neurochem 60(3):1155–1158

Querfurth HW, Selkoe DJ (1994) Calcium ionophore increases amyloid beta peptide production by cultured cells. Biochemistry 33:4550–4561

Ragno M, Tournier-Lasserve E, Fiori M, Manca A (1995) An Italian kindred with cerebral autosomal dominat arteriopathy with subcortical infarcts and leukoencephalopathy (CADASIL). Ann Neurol 38:231–236

Rebeck GW, Reiter JS, Strickland DK, Hyman BT (1993) Apolipoprotein E in sporadic Alzheimer's disease: allelic variation and receptor interactions. Neuron 11:575–580

Rebeck GW, Harr SD, Strickland DK, Hyman BT (1995) Multiple, diverse senile plaque-associated proteins are ligands of anapolipoprotein E receptor, the alpha 2-macroglobulinreceptor/low-density-lipoprotein receptor-related protein. Ann Neurol 37:211–217

Rezek DL, Morris JC, Fulling KH, Gado MH (1987) Periventricular white matter lucencies in senile dementia of the Alzheimer type and in normal aging. Neurology 37:1365–1368

Roberts GW, Lofthouse R, Allsop D, Landon M, Kidd M, Prusiner SB, Crow TJ (1988) CNS amyloid proteins in neurodegenerative diseases. Neurology 38:1534–1540

Roberts GW, Gentleman SM, Lynch A, Graham DI (1991) Beta A4 amyloid protein deposition in brain after head trauma. Lancet 338:1422–1423

Robison PM, Clemens JA, Smalstig EB, Stephenson D, May PC (1993) Decrease in amyloid precursor protein precedes hippocampal degeneration in rat brain following transient global ischemia. Brain Res 608:334–337

Roher AE, Lowenson JD, Clarke S, Woods AS, Cotter RJ, Gowing E, Ball MJ (1993) Beta-amyloid-(1–42) is a major component of cerebrovascular amyloid deposits: implications for the pathology of Alzheimer disease. Proc Natl Acad Sci USA 90:10836–10840

Rosenblum WI, Haider A (1988) Negative correlations between parenchymal amyloid and vascular amyloid in hippocampus. Am J Pathol 130:532–536

Roses AD, Saunders AM, Alberts MA, Strittmatter WJ, Schmechel D, Gorder E (1995) Apolipoprotein E E4 allele and risk of dementia. JAMA 273:374–375

Rozemuller JM, Eikelenboom P, Kamphorst W, Stam FC (1988) Lack of evidence for dysfunction of the blood-brain barrier in Alzheimer'sdisease: an immunohistochemical study. Neurobiol Aging 9:383–391

Salbaum JM, Weidemann A, Lemaire HG, Masters CL, Beyreuther K (1988) The promoter of Alzheimer's disease amyloid A4 precursor gene. EMBO J 7:2807–2813

Saunders AM, Strittmatter WJ, Schmechel D, George Hyslop PH, Joo SH, Rosi BL, Gusella JF, Crapper MacLachlan DR, Alberts MJ, Hulette C, Crain B, Goldgaber D, Roses AD (1993) Association of apolipoprotein E allele epsilon 4 with late-onset familial and sporadic Alzheimer's disease. Neurology 43:1467–1472

Scheibel AB, Duong TH, Jacobs R (1989) Alzheimer's disease as a capillary dementia. Ann Med 21:103–107

Scheltens P, Barkhof F, Leys D, Wolters EC, Ravid R, Kamphorst W (1995) Histopathologic correlates of white matter changes on MRI in Alzheimer's disease and normal aging. Neurology 45:883–888

Schmechel DE, Saunders AM, Strittmatter WJ, Crain BJ, Hulette CM, Joo SH, Goldgaber D, Roses AD (1993) Increased amyloid beta-peptide deposition in cerebral cortex as a consequence of apolipoprotein E genotype in late-onset Alzheimer disease. Proc Natl Acad Sci USA 90:9649–9653

Schneider JA, Gearing M, Robbins RS, de Mirra ASS (1995) Apolipoprotein E genotype in diverse neurodegenerative disorders. Ann Neurol 38:131–135

Scholz W (1938) Studien zur Pathologie der Hirngefässe II. Die drusige Entartung der Hirnarterien und -capillaren. Z Ges Neurol Psychiatr 162:694–715

Schubert D, LaCorbiere M, Saitoh T, Cole G (1989) Characterization of an amyloid beta precursor protein that binds heparin and contains tyrosine sulfate. Proc Natl Acad Sci USA 86:2066–2069

Schwarzman AL, Gregori L, Vitek MP, Lyubski S et al (1994) Transthyretin sequesters amyloid beta protein and prevents amyloid formation Proc Natl Acad Sci USA91:8368–8372

Shigematsu K, McGeer PL (1992) Accumulation of amyloid precursor protein in neurons after intraventricular injection of colchicine. Am J Pathol 140:787–794

Shirahama T, Skinner M, Cohen AS (1981) Immunocytochemical identification of amyloid in formalin-fixed paraffin sections. Histochemistry 72:161–171

Snow AD, Sekiguchi R, Nochlin D, Fraser P, Kimata K, Mizutani A, Arai M, Morgan DG (1994) An important role of heparan sulfate proteoglycan (Perlecan) in a mode system for the deposition and persistence of fibrillar A beta-amyloid in ratbrain. Neuron 12:219–234

Sola C, Garcia Ladona FJ, Mengod G, Probst A, Frey P, Palacios JM (1993) Increased levels of the Kunitz protease inhibitor-containing beta APP mRNAs in rat brain following neurotoxic damage. Brain Res Mol Brain Res 17:41–52

St. Clair D, Norrman J, Perry R, Yates C, Wilcock G, Brookes A (1994) Apolipoprotein E epsilon 4 allele frequency in patients with Lewy body dementia, Alzheimer's disease and age-matched controls. Neurosci Lett 176:45–46

Stephenson DT, Rash K, Clemens JA (1992) Amyloid precursor protein accumulates in regions of neurodegeneration following focal cerebral ischemia in the rat. Brain Res 593:128–135

Stewart PA, Hayakawa K, Akers MA, Vinters HV (1992) A morphometric study of the blood-brain barrier in Alzheimer's disease. Lab Invest 67:734–742

Stoll G, Müller HW (1986) Macrophages in the peripheral nervous system and astroglia in the central nervous system of rat commonly express apolipoprotein E during development but differ in their response to injury. Neurosci Lett 72:233–238

Stoll G, Müller HW, Trapp BD, Griffin JW (1989) Oligodendrocytes but not astrocytes express apolipoprotein E after injury of rat optic nerve. Glia 2:170–176

Suzuki N, Cheung TT, Cai XD, Odaka A, Otvos L, Jr., Eckman C, Golde TE (1994) An increased percentage of long amyloid beta protein secreted by familial amyloid beta protein precursor (beta APP717) mutants. Science 264:1336–1340

Tabaton M, Caponnetto C, Mancardi G, Loeb C (1991) Amyloid beta protein deposition in brains from elderly subjects with leukoaraiosis. J Neurol Sci 106:123–127

Tanaka S, Nakamura S, Ueda K, Kameyama M, Shiojiri S, Takahashi Y, Ito H (1988) Three types of amyloid protein precursor mRNA in human brain: their differential expression in Alzheimer's disease. Biochem Biophys Res Commun 157:472–479

Tanaka S, Shiojiri S, Takahashi Y, Kitaguchi N, Ito H, Kameyama M, Nakamura S, Ueda K (1989) Tissue-specific expression of three types of beta-protein precursor mRNA:enhancement of protease inhibitor-harboring types in Alzheimer's disease brain. Biochem Biophys Res Commun 165:1406–1414

Tanzi RE, Gusella JF, Watkins PC, Bruns GA, St George Hyslop P, Patterson D, Pagan S, Kurnit DM, Neve RL (1987) Amyloid beta protein gene: cDNA, mRNA distribution, and genetic linkage near the Alzheimer locus. Science 235:880–884

Tennent GA, Lovat LB, Pepys MB (1995) Serum amyloid P component prevents proteolysis of the amyloid fibrils ofAlzheimer disease and systemic amyloidosis. Proc Natl Acad Sci USA 92:4299–4303

Terry RD, Hansen LA, DeTeresa R, Davies P, Tobias H, Katzman R (1987) Senile dementia of the Alzheimer type without neocortical neurofibrillary tangles. J Neuropathol Exp Neurol 46(3):262–268

Timmers WF, Tagliavini F, Haan J, Frangione B (1990) Parenchymal preamyloid and amyloid deposits in the brains of patients with hereditary cerebral hemorrhage with amyloidosis – Dutch type. Neurosci Lett 118:223–226

Tomimoto H, Wakita H, Akiguchi I, Nakamura S, Kimura J (1994) Temporal profiles of accumulation of amyloid beta/A4 protein precursor inthe gerbil after graded ischemic stress. J Cereb Blood Flow Metab 14:565–573

Tomimoto H, Akiguchi I, Wakita H, Nakamura S, Kimura J (1995) Ultrastructural localization of amyloid protein precursor in the normal and postischemic gerbil brain. Brain Res 672:187–195

van Bockxmeer FM, Mamotte CD (1992) Apolipoprotein epsilon 4 homozygosity in young men with coronary heart disease. Lancet 340:879–880

van Duinen SG, Castano EM, Prelli F, Bots GT, Luyendijk W, Frangione B (1987) Hereditary cerebral hemorrhage with amyloidosis in patients of Dutch origins related to Alzheimer disease. Proc Natl Acad Sci USA 84:5991–5994

Vinters HV (1987) Cerebral amyloid angiopathy. A critical review. Stroke 18:311–324

Vinters HV (1991) Relationship of amyloid to Alzheimer's disease. West J Med 154(1)94–95

Vinters HV, Secor DL, Pardridge WM, Gray F (1990) Immunohistochemical study of cerebral amyloid angiopathy. III. Widespread Alzheimer A4 peptide in cerebral microvessel walls colocalizes with gamma tracein patients with leukoencephalopathy. Ann Neurol 28:34–42

Wakita H, Tomimoto H, Akiguchi I, Ohnishi K, Nakamura S, Kimura J (1992) Regional accumulation of amyloid beta/A4 protein precursor in the gerbil brain following transient cerebral ischemia. Neurosci Lett 146:135–138

Waldemar G, Christiansen P, Larsson HB, Hogh P, Laursen H, Lassen NA (1994) White matter magnetic resonance hyperintensities in dementia of the Alzheimer type: morphological and regional cerebral blood flow correlates. J Neurol Neurosurg Psychiatry 57:1458–1465

Wallace W, Ahlers ST, Gotlib J, Bragin V, Sugar J, Gluck R, Shea PA, Harou-
tunian V (1993) Amyloid precursor protein in the cerebral cortex is rapidly
and persistently induced by loss of subcortical innervation. Proc Natl Acad
Sci USA 90:8712–8716

Weisgraber KH, Roses AD, Strittmatter WJ (1994) The role of apolipoprotein
E in the nervous system. Curr Opin Lipidol 5:110–116

Wisniewski T, Ghiso J, Frangione B (1991) Peptides homologous to the amyloid protein of Alzheimer's disease containing a glutamine for glutamic
acid substitution have accelerated amyloid fibril formation. Biochem
Biophys Res Commun 180:1528

Wisniewski T, Castano EM, Golabek A, Vogel T, Frangione B (1994) Acceleration of Alzheimer's fibril formation by apolipoprotein E in vitro. Am
J Pathol 145:1030–1035

Yamada T, Kondo A, Takamatsu J, Tateishi J, Goto I (1995) Apolipoprotein E
mRNA in the brains of patients with Alzeimer's disease. J Neurol Sci
129(1):56–61

Yamaguchi H, Ishiguro K, Sugihara S, Nakazato Y, Kawarabayashi T, Sun X
(1994) Presence of apolipoprotein E on extracellular neurofibrillary tangles
and on meningeal blood vessels precedes the Alzheimer beta-amyloid deposition. Acta Neuropathol (Berl) 88:413–419

Yankner BA, Duffy LK, Kirschner DA (1990) Neurotrophic and neurotoxic
effects of amyloid beta protein: reversal by tachykinin neuropeptides.
Science 250:279–282

Zhang SH, Reddick RL, Piedrahita JA, Maeda N (1992) Spontaneous hypercholesterolemia and arterial lesions in mice lacking apolipoprotein E.
Science 258:468–471

Zlokovic BV, Ghiso J, Mackic JB, McComb JG, Weiss MH, Frangione B
(1993) Blood-brain barrier transport of circulating Alzheimer's amyloid
beta. Biochem Biophys Res Commun 197:1034–1040

Zlokovic BV, Martel CL, Mackic JB, Matsubara E, Wisniewski T, McComb
JG, Ghiso J (1994) Brain uptake of circulating apolipoproteins J and E
complexed to Alzheimer's amyloid beta. Biochem Biophys Res Commun
205:1431–1437

Zlokovic BV, Martel CL, Matsubara E, Frangione G, Ghiso J (1995) Cellular
uptake of apolipoprotein J and Alzheimer's amyloid beta complexed to
apolipoprotein J at the blood-brain barrier and in the choroid plexus Soc
Neurosci Abstr 21:6 (abstr)

3 The Genes Involved in Alzheimer's Disease

J. Hardy and M. Hutton

3.1 Introduction

So far, four genes have been shown to be involved in the etiology of Alzheimer's disease (AD). These are the β-amyloid precursor protein (APP) gene on chromosome 21 (the AD1 locus: Goate et al. 1991), the apolipoprotein E (ApoE) gene on chromosome 19 (the AD2 locus: Corder et al. 1993), the presenilin 1 (PS-1) gene on chromosome 14 (the AD3 locus: Sherrington et al. 1995), and the presenilin 2 (PS-2) gene on chromosome 1 (AD4 locus: Levy-Lahad et al. 1995). In addition, the α1-antichymotrypsin (AACT) gene, also on chromosome 14, has been proposed as another locus involved in the disease (Kamboh et al. 1995), but this awaits confirmation.

This review will focus largely on the PS-1 and PS-2 genes, with a small section on APP, for a number of reasons. First, these loci act as simple autosomal dominant loci, whereas both ApoE and possibly AACT appear to behave in more complex ways. Second, it seems as if the PS and APP mutations act via related mechanisms (Younkin, unpublished data). Third, it is possible that ApoE may not have a specific role in AD, but rather, a more general role in neuronal repair (Poirier 1994); fourth, there have been numerous recent reviews on the roles of both APP and ApoE in AD (Selkoe 1994; Roses 1994; Hardy 1995).

3.2 β-Amyloid Precursor Protein

For a more extensive review of APP and its role in AD, see Selkoe (1995). Briefly, the abundant literature on APP and the genetics of AD can be summarized as follows: mutations in APP which cause AD (Goate et al. 1991; Murrell et al. 1991; Chartier Harlin et al. 1991; Mullan et al. 1992; Hendricks et al. 1992) do so because they cluster at critical positions in the APP molecule at cleavage sites (Hardy 1992: see Fig. 1). Thus, the Swedish mutation alters APP processing in such a way that more Aβ is produced (Citron et al. 1992; Cai et al. 1993), the Flemish mutation (Hendricks et al. 1992) reduces the flux of APP down the alternative pathway, also increasing Aβ production (Haass et al. 1994), and the London mutations (Goate et al. 1991; Murrell et al. 1991; Chartier Harlin et al. 1991) alter APP processing such that a greater proportion of Aβ is Aβ 1-42/43 rather than the shorter (and more soluble) Aβ 1–40 (Suzuki et al. 1994).

These AD-causing APP mutations increase the likelihood of Aβ deposition by increasing the concentration of Aβ, in particular, Aβ1-

Fig. 1. The APP mutations which cause AD have defined effects on β-amyloid ▶ production. *Panel 1* shows normal APP metabolism with both α and β pathways operating. *Panel 2* shows the effect of the APP670/1 mutation which potentiates the β pathway and leads to increased production of β-amyloid. *Panel 3* shows the effects of the APP717 mutations which lead to production of a greater proportion of longer β-amyloid. *Panel 4* shows that the effect of the APP692 (and possibly also the APP 693) mutation is to inhibit the α pathway and cause a greater proportion of APP to be processed by a β-like pathway

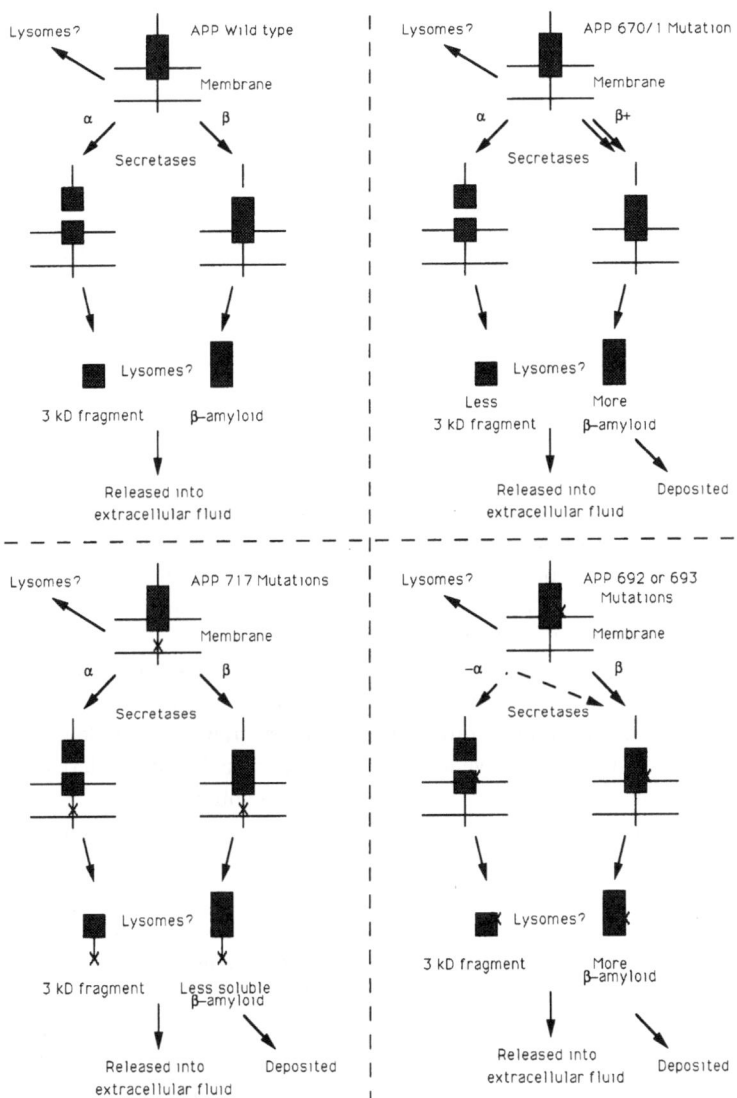

Fig. 1. Legend see p. 50

42/43. The APP mutation of APP, which causes the related disease, hereditary cerebral hemorrhage with amyloidosis (Dutch type) (Levy et al. 1990), acts via a slightly different mechanism of reducing the solubility of Aβ 1–40 (Wisniewski et al. 1991). This work is summarized schematically in Fig. 1.

Thus, APP mutations are dominant, gain of function changes, where the gain of function is Aβ deposition. The age of onset of disease in families with APP mutations is typically around 48–60 years (Houlden et al. 1993).

3.3 Presenilin 1 Gene

3.3.1 Cloning and Structure

The PS-1 gene was cloned by positional cloning strategies (Sherrington et al. 1995). It was an anonymously expressed sequence cloned from the candidate region on chromosome 14 (Sherrington et al. 1995) and had no homologies to proteins of known function or structure (but see below). Because the structure of the protein is derived entirely from its hydropathy profile, it only represents a "best guess" (see Fig. 2). Based on the hydropathy profile, there appear to be seven transmembrane domains; however, other protein prediction programs and window sizes suggest anything from five to nine transmembrane domains (Clark et al. 1995). There is no homology to seven transmembrane domain, G-protein-coupled receptors. The open reading frame of the gene is encoded by ten exons and there is some alternate splicing (Rogaev et al. 1995; Clark et al. 1995). Two regions in particular show alternate splicing: a VRSQ motif at codons 26–29 in the N-terminal domain can be spliced in or out, depending on different splice donor site usage in exon 3, and exon 8 can be spliced out from the long loop between transmembrane domains 6 and 7 (reported in Rogaev et al. 1995). The functional significance of the alternate splicing is not known, although the VRSQ is a potential phosphorylation site for protein kinase C and casein kinase II and, thus, this splicing might have regulatory significance (Clark et al. 1995).

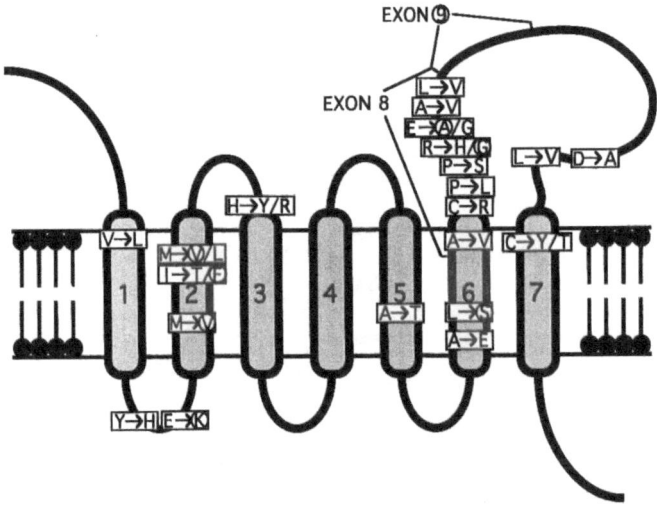

Fig. 2. The proposed structure of presenilin-1 (PS-1; PS-2 is similar) showing the seven putative transmembrane domains. The N terminal is probably cytoplasmic. The positions of mutations are also shown (see Van Broeckhoven 1995)

3.3.2 Mutations

A large number of mutations which lead to early onset of AD have been described in the PS-1 gene (Sherrington et al. 1995; Wasco et al. 1995; Clark et al. 1995; Van Broeckhoven 1995). It is noteworthy that, so far, no nonsense or frameshift mutations have been found: all the mutations are missense, conservative changes to residues which are conserved between both PS-1 and PS-2 (below). This absence of destructive mutations suggests, but does not prove, that the mutations act by a gain of function since the most effective way to reduce function would be to eliminate one allele completely by a nonsense or frameshift change. This logic is compelling, but not unassailable, since it is possible that through protein dimerization, the mutant allele can inactivate the wild-type allele and reduce the PS activity to less than the 50% of a single allele. It is also possible that nonsense or frameshift changes will be found. However, the simplest explanation is that the mutations act by a

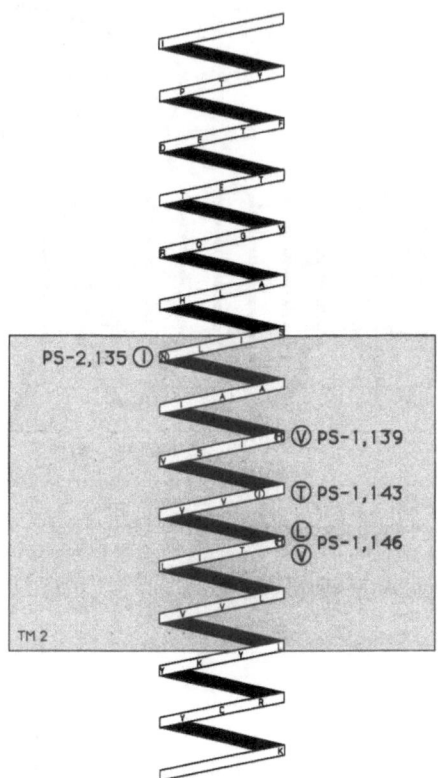

Fig. 3. The structure of the proposed transmembrane domain 2 of the presenilin 1 protein showing that the presenilin 1 mutations at codons 139, 143, and 146 line up on one side of the α-helix. The presenilin 2 mutation at codon 141, for which the homologous presenilin 1 codon is 135, probably also fits with this cluster because it is located one turn of the helix up from codon 139. The clustering of these mutations suggests that this surface of transmembrane domain 2 is important, for example, lining a channel or interfacing with another protein, and when its function is disrupted AD develops

gain of a deleterious function. This gain may be of a new function or, more likely perhaps, by amplification of a preexisting function (as was the case with APP mutations outlined above).

The mutations appear to group in three clusters. These clusters are around transmembrane domain 2 (cluster 1), centering around the prolines at around codon 265 (cluster 2) and at around codon 400 (cluster 3). It is difficult to discern an underlying logic yet to clusters 2 and 3 but some of the mutations in cluster 1 appear to line up on one side of the putative α-helix as it passes through the membrane (Fig. 3). In this position, they might conceivably disrupt the interaction between PS-1 and another membrane protein, or perhaps, they may alter the ion gating properties if the PS proteins are ion channels (Sherrington et al. 1995; Clark et al. 1995). Mutations in the PS-1 gene lead to disease with an onset age between 30 and 55 years (Houlden et al. 1993; Sherrington et al. 1995).

3.4 Presenilin 2 Gene

3.4.1 Cloning and Structure

The PS-2 gene was cloned because of its homology with the PS-1 gene. When the former was cloned, the screening of public gene sequences databases revealed a series of ESTs (expressed sequence tags) which each encoded a short segment of sequence which was homologous to PS-1. Library screening revealed that these short sequences all came from a single isologue of PS-1 which had 65% homology and mapped to chromosome 1 (Levy-Lahad et al. 1995a; Clark et al. 1995; Rogaev et al. 1995). Independently, the locus leading to the Volga German variant of familial AD had been mapped to chromosome 1 (Levy-Lahad et al. 1995b) and sequencing the locus in Volga German families revealed a mutation in this sequence (see below).

The splicing pattern of the PS-2 gene is considerably more complex than that of the PS-1 gene and there is evidence for considerable alternative splicing (Levy-Lahad et al. 1995a; Rogaev et al. 1995), although it is clear that the two genes are homologous, not only with respect to their open reading frame, but also because the exon boundaries appear to be similar (Clark et al. 1995 and unpublished).

3.4.2 Mutations

So far, only two mutations have been described in PS-2, the original one found in the Volga Germans, where it represents a founder effect, and a second, found in an Italian pedigree. Both mutations fit the "rule" that they alter residues conserved between PS-1 and PS-2, and the Volga German mutation fits in cluster 1 in transmembrane domain 2 where it alters the residue homologous to N135, four residues (one helix turn) above M139. Although only two mutations have been found so far in PS-2, there is every reason to suppose that there will be as many as there are in PS-1, but the PS-2 mutations appear (on the basis of very limited information) to lead to a more variable, and generally older onset age than the PS-1 changes (Levy-Lahad et al. 1995b; Rogaev et al. 1995). Families with variable onset age will be ascertained less frequently than those with consistent and early onset ages, so it may prove to be more difficult to find PS-2 mutations because fewer samples will have been collected.

3.5 Functions and Dysfunction of PS

There are two clues, both from *Caenorhabditis elegans* work, about the functions of the PSs. The first is that elimination of one isologue of the PSs prevents sperm maturation by disrupting the formation of the equivalent of the Golgi apparatus (L'Hernault and Arduengo 1992). The second is that disruption of another isologue disrupts Notch signaling (Levitan and Greenwald 1995). The significance of these observations to the PSs remains to be established but *C. elegans* offers an immensely powerful system for the genetic dissection of the function of their isologues.

It may be relevant in this regard that the Golgi apparatus is critical to the correct processing of APP (Xui et al. 1995) and this might supply the link between the physiology of these molecules and their pathology. Of particular interest is the observation by Younkin and colleagues (unpublished) that fibroblasts from carriers of both PS-1 and PS-2 mutations make more $A\beta$ 1–42/43 in a way reminiscent of the APP London mutations.

Thus, while it is much too early to be certain, the major hypothesis for now seems to be that PS mutations lead to disease by subtle disruption of APP processing, perhaps in the Golgi formation.

References

Cai XD, Golde TE, Younkin SG (1993) Excess amyloid beta-protein is released from a mutant amyloid beta-protein precursor linked to familial Alzheimer's disease. Science 259:514–516

Chartier Harlin MC, Hardy J, Mullan M (1991) Early onset Alzheimer's disease caused by mutations at codon 717 of the β-amyloid precursor protein gene. Nature 353:844–846

Citron M, Olterdorf T, Haass C, McConlogue L, Hung AY, Seubert P, Vigo-Pelfrey C, Lieberburg I, Selkoe DJ (1992) Mutation of the beta-amyloid precursor protein in familial Alzheimer's disease causes increased beta-amyloid production. Nature 360:672–674

Clark R, Hutton M, Fuldner R et al (1995) The structure of the presenilin I (S182) gene and identification of six novel mutations in early onset Alzheimer's disease families. Nature Genet 11:219–222

Corder EH, Saunders AM, Strittmatter WJ, Schmechel DE, Gaskell PC, Small GW, Roses AD, Haines JL, Pericak-Vance MA (1993) Gene dose of apolipoprotein E type 4 allele and the risk of Alzheimer's disease in late onset families. Science. 261:921–923

Goate A, Chartier-Harlin MC, Mullan M, Brown J, Crawford F, Fidani L, Giuffra L, Haynes A, Irving N, James L, Mant R, Newton P, Rooke K, Roques P, Talbot C, Pericak-Vance M, Roses A, Williamson R, Rossor M, Owen M, Hardy J (1991) Segregation of a missense mutation in the amyloid precursor protein gene with familial Alzheimer's disease. Nature 349:704–706

Haass C, Hung AY, Selkoe DJ, Teplow DB (1994) Mutations associated with a locus for familial Alzheimer's disease result in alternative processing of amyloid P protein precursor. J Biol Chem 269:17741–17748

Hardy JA (1992) Framing β-amyloid. Nature Genet 1:233–234

Hardy J (1995) Apolipoprotein E in the genetics and epidemiology of Alzheimer's disease. J Neuropsychiatr Genet 60:456–460

Hendricks L, van Duijn CM, Cras P, Cruts M, Van Hul W, van Harskanp F, Warren A, McInnis MG, Antonarakis SE, Martin JJ, Hofman A, Van Broeckhoven C (1992) Presenile dementia and cerebral haemorrhage caused by a mutation at codon 692 of the beta-amyloid precursor protein gene. Nature Genet 1:218–221

Houlden H, Collinge J, Kennedy A, Newman S, Rossor M, Lannfelt L, Lilius L, Winblad B, Crook R, Duff K, Hardy J (1993) ApoE genotype and Alzheimer's disease. Lancet 342:737–738

Kamboh MI, Sanghera DK, Ferrell RE, DeKosky ST (1995) APOE4 associated Alzheimer's disease risk is modified by α1 antichymotrypsin polymorphism. Nature Genet 10:486–488

Levitan D, Greenwald L (1995) Facilitation of lin-12 mediated signalling by sel-12, a Caenorhabditis elegans S182 Alzheimer's disease gene. Nature 377:351–354

Levy E, Carman MD, Fernandez-Madrid IJ, Power MD, Lieberburg I, van Duinen SG, Bots GTAM, Luyendijk W, Frangione B (1990) Mutation of the Alzheimer's disease amyloid gene in hereditary cerebral hemorrhage, Dutch type. Science 248:1124–1126

Levy-Lahad E, Wasco W, Poorkaj P, Romano DM, Oshima J, Pettingell WH, Yu CE, Jondro PD, Schmidt SD, Wang K et al. (1995a) Candidate locus for the chromosome 1 familial Alzheimer's disease locus. Science 268:973–977

Levy-Lahad E, Wijsman EM, Nemens E, Anderson L, Goddard KA, Weber JL, Bird TD, Schellenberg GD (1995b) A familial Alzheimer's disease locus on chromosome 1. Science 268:970–972

L'Hernault SW, Arduengo PM (1992) Mutation of a putative sperm membrane protein in *Caenorhabditis elegans* prevents sperm differentiation but not its associated meiotic divisions. J Cell Biol 119:55–68

Mullan M, Crawford F, Axelman K, Houlden H, Lilius L, Winblad B, Lannfelt L (1992) A new mutation in APP demonstrates that pathogenic mutations for probable Alzheimer's disease frame the beta-amyloid sequence. Nature Genet 1:345–347

Murrell J, Farlow M, Ghetti B, Benson MD (1991) A mutation in the amyloid precursor protein associated with hereditary Alzheimer's disease. Science 254:97–99

Poirier J (1994) Apolipoprotein E in animal models of CNS injury and in Alzheimer's disease. Trends Neurosci 17:529–530

Rogaev EI, Sherrington R, Rogaeva EA, Levesque G, Ikeda M, Liang Y, Chi H, Lin C, Holman K, Tsuda T et al (1995) Familial Alzheimer's disease in kindreds with missense mutations in a gene on chromosome 1 related to the Alzheimer's disease type 3 gene. Nature 376:775–778

Roses AD (1994) Apolipoprotein E affects the rate of Alzheimer's disease expression: beta-amyloid burden is a secondary consequence dependent on ApoE genotype and duration of disease. J Neuropathol Exp Neurol 5:429–437

Selkoe DJ (1994) Alzheimer's disease: a central role for amyloid. J Neuropathol Exp Neurol 5:438–447

Sherrington R, Rogaev EI, Liang Y, Rogaeva EA, Levesque G, Ikeda M, Chi
    H, Lin C, Li G, Holman K et al. (1995) Cloning of a gene bearing missense
    mutations in early onset Alzheimer's disease. Nature 375:754–760
Suzuki N, Cheung TT, Cai XD, Odaka A, Otvos L Jr, Eckman C, Golde TE,
    Younkin SG (1994) An increased percentage of long amyloid beta-protein
    secreted by familial amyloid beta-protein precursor (APP717) mutants.
    Science 264:1336–1340
Van Broeckhoven CM (1995) The presenilins in Alzheimer's disease. Nature
    Genet 11:230–232
Wasco W, Pettingell WP, Jondro PD, Schmidt SD, Gurubhagavatula S, Rodes
    L, Diblasi T, Romano DM, Guenette SY, Kovacs DM, Growdon JH, Tanzi
    RE (1995) Familial Alzheimer's disease chrmsome 14 mutations. Nature
    Med 11:848
Wisniewski T, Ghiso J, Frangione B (1991) Peptides homologous to the amy-
    loid protein of Alzheimer's disease containing a glutamine for glutamate
    substitution have accelerated fibril formation. Biochem Biophys Res Com-
    mun 179:1247–1254
Xui H, Greengard P, Gandy S (1995) Regulated formation of Golgi secretory
    vesicles containing Alzheimer β-amyloid precursor protein. J Biol Chem 40
    (in press)

4 In Vivo Biology of Amyloid Precursor Protein/Amyloid Precursor-like Proteins and Transgenic Animal Models of Alzheimer's Disease

S.S. Sisodia, G. Thinakaran, B.T. Lamb, H.H. Slunt,
C.S. von Koch, S.D. Ginsberg, A.C.Y. Lo, M.K. Lee,
A.J.I. Roskams, E. Masliah, H. Zheng, L.H.T. Van der Ploeg,
J.D. Gearhart, and D.L. Price

4.1 Biology of Amyloid Precursor Protein

A principal pathological hallmark of Alzheimer's disease (AD) is the presence of extracellular deposits of β-amyloid protein (Aβ), a 39–43 amino acid peptide comprised of 11–15 amino acids of the transmembrane domain and 28 amino acids of the extracellular domain of amyloid precursor protein (APP; Glenner and Wong 1984; Masters et al. 1985a, b). Aβ is associated with neurites in the classical senile plaque (Wisniewski and Terry 1973; Glenner and Wong 1984; Masters et al. 1985a, b; Selkoe et al. 1987; Probst et al. 1991), with preamyloid deposits, and with amyloid within the walls of leptomeningeal and cerebral vessels (Kemper 1984; Selkoe et al. 1987).

APP, encoded by alternatively spliced mRNA, mature through the constitutive secretory pathway and are modified by the addition of both N- and O-linked carbohydrates, phosphate, and sulfate moieties (Oltersdorf et al. 1989; Weidemann et al. 1989; Hung and Selkoe 1994). Varying levels of newly synthesized APP molecules appear at the cell surface (Weidemann et al. 1989; Haass et al. 1992a; Sisodia 1992); some of these molecules are cleaved endoproteolytically by APP "α-secretase" within the Aβ sequence (Esch et al. 1990; Sisodia et al. 1990; Anderson et al. 1991; Wang et al. 1991) to release the ectodomain of APP, including residues 1–16 of Aβ, into the medium. Thus, APP cleavage within the Aβ domain precludes the formation of Aβ. A fraction of cell-surface APP is also internalized and degraded via endosomal–lysosomal pathways (Cole et al. 1989; Golde et al. 1992; Haass et al. 1992a). Processing via the endosomal–lysosomal pathway results in the production of fragments that contain the entire Aβ region and APP C-terminus and are, hence, potentially amyloidogenic (Golde et al. 1992; Haass et al. 1992a). However, several lines of evidence now suggest that the lysosomal degradation of APP is unlikely to contribute to the production of Aβ (reviewed in Haass and Selkoe 1993). It is now clear that Aβ are secreted constitutively by primary and tissue culture cells (Haass et al. 1992b; Shoji et al. 1992; Busciglio et al. 1993) and are present in cerebrospinal fluid (Shoji et al. 1992; Seubert et al. 1993). Although the cellular and molecular mechanisms of Aβ generation are not clearly understood, several recent studies have begun to clarify these processes. For example, agents that interfere with pH gradients (i.e., ammonium chloride and chloroquine) inhibit the production of Aβ (Haass et al. 1992b; Shoji et al. 1992), suggesting that Aβ may be generated in acidic compartments (i.e., endosomes or late Golgi cells). In support of this model, studies have demonstrated that Aβ can be generated from surface-labeled APP (Koo and Squazzo 1994). Although these studies suggest that Aβ may be produced and released in vitro and in vivo, the relationships of these Aβ-related fragments to deposits of Aβ in AD (principally 42–43 amino acids) have not been fully defined.

The demonstration of autosomal dominant linkage of several missense mutations of APP in a relatively small subset of patients with early-onset familial AD has provided unambiguous support to the view that APP/Aβ are central to the etiopathogenesis of disease in these individuals (Goate et al. 1991; Mullan et al. 1992). In ~12 early-onset pedigrees, missense mutations generated amino acid substitutions at residue 717 (of APP-770) within the transmembrane domain of APP (Chartier-Harlin et al. 1991; Goate et al. 1991; Naruse et al. 1991). In one family with a mutation at codon 692 of APP, which resulted in a Gly-Ala substitution, biopsies from affected individuals disclosed diffuse deposits of Aβ, congophilic angiopathy, and scattered senile plaques but no neurofibrillary tangles (NFT; Hendricks et al. 1992). Finally, in two large, related families from Sweden (Mullan et al. 1992), a double mutation at codons 670 and 671 that resulted in the substitution of Lys-Met to Asn-Leu was linked to early-onset AD. Cellular transfection approaches have provided considerable insight regarding the mechanism whereby mutations in APP affect the processing of the precursor protein and contribute to Aβ production. For example, tissue culture cells expressing APP harboring the "Swedish" substitutions secrete higher levels of Aβ–containing peptides as compared to cells expressing wild-type constructs (Citron et al. 1992; Cai et al. 1993). Additional studies of cells that express APP harboring the Ala–Gly substitution at amino acid 692 (Hendricks et al. 1992) reveal that α-secretase processing of this mutant polypeptide is inefficient and that secreted Aβ species exhibit considerable microheterogeneity, including the appearance of more hydrophic species (Haass et al. 1994). Finally, studies have demonstrated that cells expressing APP harboring "717" substitutions do not appear to secrete higher levels of Aβ but rather secrete a higher fraction of longer Aβ peptides (i.e., extending to Aβ residue 42) relative to cells that express wild-type APP (Suzuki et al. 1994). The finding that APP harboring 717 mutations is processed to generate higher levels of longer Aβ peptides is of considerable interest, because recent physicochemical studies have indicated that amyloid formation is a nucleation-dependent phenomenon and that the C-terminus may be a critical determinant of the rate of amyloid formation (Jarrett et al. 1993; Jarrett and Lansbury 1993). In support of the latter physicochemical studies, elegant immunocytochemical studies with antibodies uniquely specific for Aβ1–40 or Aβ1–42 revealed that Aβ deposition most likely begins with Aβ1–42 (43) and not with Aβ1–40 (Iwatsubo et al. 1994). Ultimately, it will be essential to generate transgenic animals which express APP variants harboring missense substitutions linked to familial AD (FAD) in order to recapitulate neuropathological features of AD (see Sect. 4.4).

The biological functions of APP in the central nervous system (CNS) have not been fully established, although APP transcripts/proteins are expressed in most neurons and nonneuronal cells (Higgins et al. 1988; Neve et al. 1988; Palmert et al. 1988; Shivers et al. 1988; Koo et al. 1990b; Cras et al. 1991; Haass et al. 1991; Martin et al. 1991). In the rat peripheral nervous system, holo-APP-695 synthesized by neurons is carried by fast anterograde axonal transport (Koo et al. 1990a; Sisodia et al. 1993) and presumably delivered to nerve terminals. Similarly, APP is transported in CNS pathways and appear to be proteolytically cleaved at nerve terminals by mechanisms which are presently unclear (S.S. Sisodia, V.E. Koliatsos, J. Buxbaum, and D.L. Price, personal observations). Other studies have indirectly suggested that APP play roles in cell–cell or cell–matrix interactions (Shivers et al. 1988; Schubert et al. 1989; Klier et al. 1990), calcium hemostasis (Mattson et al. 1993a), growth-promoting activities (Saitoh et al. 1989), and the formation/maintenance of synapses in vivo (Mucke et al. 1994). More recently, investigations in cultured cells (Mattson et al. 1993b) and in transgenic animals (Mucke et al. 1995) have provided support to the notion that APP is neuroprotective.

4.2 Gene Targeting of *APP*

In an attempt to assess APP function in vivo, we used a gene-targeting strategy to generate mice with functionally inactivated *APP* genes (Zheng et al. 1995). Mice with a homozygous mutation in the *APP* gene are viable but exhibit reactive gliosis in the cortex and hippocampus and decreased locomotor activity. Furthermore, we have examined the expression of mRNAs encoding each of the APP homologues in mice with inactivated APP genes and have failed to demonstrate any compensatory changes in the levels of amyloid precursor-like protein (APLP) 1 or 2 transcripts or polypeptides. Moreover, we were unable to detect any alterations either at the level of cellular morphology or cortical organization in the CNS of *APP*-targeted mice relative to wild-type littermates. Finally, in an attempt to ascertain a potential functional role for APP in neuronal sprouting/synaptogenesis, we injured the perforant pathway by aspirating the entorhinal cortex. Two weeks later, we failed to observe any differences between *APP*-targeted and wild-type mice in the level of reactive synaptogenesis in the outer molecular layers of the dentate gyrus, as assayed by acetylcholinesterase histochemistry.

4.3 Biology of APLPs

APLP1, the APP homologue first to be described, was shown to be expressed specifically and at high levels in adult brain (Wasco et al. 1992). We have raised polyclonal antibodies against a synthetic peptide corresponding to the C-terminal 11 amino acids of APLP1 and document that APLP1 is uniquely expressed in mouse brain and that there is little selectivity between brain regions. We have analyzed the developmental expression of *APLP1* in mouse by in situ hybridization and document that the first detectable expression of APLP1 transcripts is coincident with the onset of neurogenesis (days 9–10 postimplantation) with newly born, postmitotic neurons in the developing CNS and in the spinal cord expressing abundant APLP1 mRNA. The pattern of expression of APLP1 mRNA is distinct from the generally ubiquitous distributions of APP and APLP2 transcripts at these early developmental stages. Furthermore, we have used reverse transciptse polymerase chain reaction (RT-PCR) to examine the expression of APLP1 mRNA in a pluripotent mouse cell line, P19, stimulated to differentiate by treatment with retinoic acid. We show that the appearance of APLP1 mRNA parallels the expression of neurofilament mRNA. Interestingly, we have not detected any APLP1 transcripts in differentiating P19 cells or in mouse development that encode the Kunitz protease inhibitor (KPI) domain. The biological function(s) of APLP1 at early stages of neuronal differentiation are unclear and await the generation and characterization of mice with targeted *APLP1* alleles.

Like APP, APLP2 is encoded by several mRNAs derived by alternative splicing (Wasco et al. 1993; Sandbrink et al. 1994; Slunt et al. 1994), and APLP2 matures through the same unusual secretory/cleavage pathway as APP (Slunt et al. 1994). However, in contrast to APP-695 and APP-751/770, APLP2-751 is a substrate for modification by a single chondroitin sulfate glycosaminoglycan (CS GAG) (Thinakaran and Sisodia 1994). Several alternatively spliced APLP2 transcripts that contain or lack sequences encoding a 12 amino acid miniexon are differentially expressed in specific neuronal populations (see below). Interestingly, encoded polypeptides that lack sequences encoded by the miniexon are substrates for modification by CS GAG chains at a single serine residue; remarkably, insertion of sequences encoded by the miniexon fully abrogates CS GAG attachment. We also demonstrate that L-APP isoforms, which lack 18 amino acids encoded by exon 15, are also substrates for CS GAG addition (Thinakaran et al. 1995). The molecular mechanisms involved in regulating the levels of alternatively spliced transcripts that encode APP/APLP2 isoforms, some of which are modified by CS GAG, are unknown. In any event, we suggest that the CS GAG addition represents a cell- or developmental-specific mechanism to generate additional functional diversity for each polypeptide.

To define the distribution of APLP2 in the rodent nervous system, we generated antibodies specific for APLP2. Our investigations with the APLP2-specific antibodies revealed that APLP2 is expressed in a variety of regions, including cerebral cortex, hippocampus, and thalamus, and that APLP2 prominently colocalizes with microtubule-associated protein 2. In contrast to its subcellular distribution in cortex and hippocampus, APLP2 is clearly present in both pre- and postsynaptic compartments in the olfactory bulb. Notably, mRNA encoding CS GAG-modified forms of APLP2 are highly represented in the olfactory epithelium, relative to alternatively spliced mRNA that encode CS GAG-free forms of APLP2. As would be anticipated, high levels of sensory neuron-synthesized CS GAG-modified APLP2 accumulates in the olfactory bulb. Because sensory neurons in the olfactory epithelium are in a state of continual turnover, axons of newly generated neurons must establish synaptic connections with neurons in the olfactory bulb in adult life. In view of the evidence in support for CS proteoglycans in aspects of cell migration and neuronal outgrowth, we suggest that CS GAG-modified APLP2 plays an important role in axonal pathfinding and/or synaptic interactions of newly born olfactory sensory neurons with respective targets in the olfactory bulb (Thinakaran et al. 1996). In parallel, we have demonstrated that APLP2 is anterogradely transported by the fast component of peripheral sensory neurons. Notably, the principal APLP2 isoform expressed and transported by sensory neurons in the dorsal root ganglia is modified by CS GAG. Since the terminal fields of sensory neurons of the olfactory epithelium and dorsal root ganglia are highly plastic structures, we suggest that the CS GAG forms of APLP2 plays an important role in sprouting and terminal remodeling. To assess the potential biological functions of APLP2 in vivo, we have used a gene targeting strategy to generate mice with inactivated *APLP2* genes. We demonstrate that, although APLP2 is absent in these animals, there are no compensatory changes in the levels of APP or APLP1 in the nervous system or systemically. Mice with inactivated *APLP2* genes will be valuable reagents for assessing the role of APLP2 in experimental paradigms of degeneration and regeneration in the CNS and the peripheral nervous system.

4.4 Transgenic Animal Models of AD

To date, the most successful studies of AD-type brain alterations have taken advantage of the finding that aged nonhuman primates develop behavioral and brain abnormalities similar to those in humans (for review, see Price and Sisodia 1994). Nevertheless, clarification of the biochemical mechanism(s) which underlie the phenotypic changes in these animals has been hampered in part by the variability in phenotype and the scarcity of aged animals. On the other hand, generation of transgenic mice which express mutated genes that are genetically linked to AD offer a unique opportunity to recapitulate aspects of the neuropathology of AD and clarify the biochemical basis for these phenotypes. Some of the efforts to produce mice with Aβ deposits using cDNA and yeast artificial chromosome (YAC)-based transgenic technologies are reviewed below.

One of the first published studies of transgenic mice were those that described mice expressing human APP-751 under the control of neuron-specific enolase promoter. Extracellular Aβ deposits and A68-immunoreactive processes were present in cortex and hippocampus (Higgins et al. 1994). However, concerns about the specificity of the immunological reagents, the lack of information on absolute levels of transgene-derived mRNA or polypeptide products, and the limited scope of the pathology has made it difficult to interpret the significance of the study (Price and Sisodia 1994). Nevertheless, this group has published a behavioral study suggesting that the 12-month-old transgene mice develop spatial learning defects (Moran et al. 1995).

The finding that several missense mutations in APP are genetically linked to pedigrees with early-onset AD has led investigators to assess the phenotype of transgenic mice that overexpress mutant APP. Dr. Hsaio and Dr. Borchelt and colleagues (Borchelt et al. 1994) have created transgenic FVB mice that express myc epitope-tagged human APP-695 (HuAPP-695myc) or HuAPP-695myc harboring the APP-717 mutation, placed under the transcriptional control of the hamster prion gene promoter. Several lines of transgenic mice were generated with these constructs. No developmental or pathological abnormalities were evident in "wild-type" animals despite abundant HuAPP-695myc expression in all neurons of the CNS; the level of total APP was elevated ~2.5-fold in wild-type lines. These animals developed behavioral disorders, including inactivity, agitation, neophobia, and seizures. Glucose utilization is diminished in cortical limbic areas, and animals die prematurely. The age of onset of illness decreases with increasing levels of brain APP. No extracellular amyloid has been detected, but there is significant gliosis. Because a similar neurological disorder develops naturally in older nontransgenic FVB mice, it has been argued that this disease may be an age- and strain-related dysfunction of the CNS exacerbated by the presence of the transgene. Mice that expressed mutant APP at levels similar to or slightly higher than wild-type expressors showed markedly reduced life spans (50–100 days). LaFerla and colleagues (1995) have generated a line of mice carrying a transgene encoding for the murine homologue of $A\beta1$–42 under the control of the 68-kDa polypeptide neurofilament promoter (NF-L). Expression of the transgene was confirmed by Northern analysis and positive in situ hybridization in hippocampus and cerebral cortex. Neuropathological examination of these animals, which suffered seizures and died at higher rates than controls, showed extensive cell death, apoptosis, and intense gliosis in cerebral cortex and hippocampus. No senile plaques were identified by silver staining, but apparent extracellular $A\beta$ immunostaining was detected in the neuropil.

Games and coworkers (1995) reported the exciting finding that transgenic mice produced by pronuclear injection express high levels of human mutant APP in which a valine 717 is substituted by phenylalanine and develop extracellular thioflavin S-positive $A\beta$ deposits as well as neurites. The construct used to generate these mice utilized the platelet-derived growth factor-β promoter driving a human APP minigene encoding the $APP_{717V\rightarrow F}$ mutation associated with FAD. The construct contained APP introns 6–8 that allow alternative splicing of exons 7 and 8. Southern blots disclosed ~40 copies of the transgene inserted at a single site and transmitted in a stable fashion. Levels of human APP mRNA and protein were significantly greater than endogenous APP transcript, and the three major splicing variants of APP were demonstrable. Significantly, levels of the transgene product were tenfold higher than endogenous mouse APP. Moreover, a 4-kDa Aβ-immunoreactive peptide was identified in the brains of these animals. By 8 months of age, Aβ deposits were seen in the hippocampus, corpus callosum, and cerebral cortex; these deposits increased in number over time. Deposits range from diffuse irregular types to compacted plaques with cores. Amyloid deposits were stained by the thioflavin, Congo red, and Bielschowsky methods. Many plaques showed glial fibrillary acid protein-positive astrocytes as well as microglial cells. Although there were distorted neurites, often present in proximity to plaques, tau-positive neurites and NFT were not present. To date, behavioral abnormalities have not been described, and no quantitative estimates of loss of neurons have been published. These mice develop Aβ amyloid deposits in the brain. This work clearly shows that it is possible to create animals with some of the abnormalities that occur in human AD.

Our efforts have focused on generating transgenic animals that express human APP encoded by the entire human *APP* gene. A YAC containing ~650 kilobases of human genomic DNA, including the *APP* gene, was transfected into embryonic stem (ES) cells (Lamb et al. 1993). ES cells that contain stably integrated YAC DNA were microinjected into mouse blastocysts, and chimeric mice were generated. After breeding, it was established that human APP sequences were transmitted to the mouse germline. Furthermore, human APP mRNA is actively transcribed in mouse tissue, and the splicing pattern of human APP transcripts in transgenic mouse tissue mirrored the endogenous pattern of alternatively spliced mRNA. Using antibodies specific for human APP, Western blot analysis of transgenic mouse brain extracts revealed that human APP contributed ~40% of total APP levels (Lamb et al. 1993). No AD-type pathology was demonstrable in animals as old as 2 years (unpublished observations). We initiated several breeding studies intended to increase human APP gene copy number and, hence, APP levels. We have failed to observe any pathological alterations in animals up to 1 year of age that express up to 2.5-fold the level of endogenous APP. In any event, these animals will be useful reagents for further studies involving breeding to the APP null mice (see Sect. 4.4). In recent studies, we have exploited the YAC–ES strategy to introduce modified human APP YAC that encode FAD mutations into the mouse germline. We have generated several lines with germline transmission of the human *APP* gene containing mutations at codon 717 singly or in combination with the Swedish double mutation located at codons 670 and 671. We anticipate that by expressing APP harboring missense mutations linked to FAD, animals will be predisposed to develop parenchymal Aβ deposits and, possibly, other brain abnormalities that occur in individuals with AD.

Finally, and in view of the exciting discoveries of autosomal dominant linkage of mutations in genes encoded on chromosome 14 (S182) (Sherrington et al. 1995) and chromosome 1 (STM2) (Levy-Lahad et al. 1995; Rogaev et al. 1995) to early- and late-onset FAD, respectively, we have begun studies aimed at generating transgenic animals expressing these mutated polypeptides.

In summary, it is critical to generate transgenic mice which express mutant APP and S182 genes in order to recapitulate the AD-type alterations in behavior and brain structure and function. These models will be invaluable for future therapeutic approaches designed to improve age-related neuropathological and behavioral deficits that occur in humans with AD.

Acknowledgments. The authors gratefully acknowledge our colleagues Dr. David Borchelt, Dr. Randall Reed, Dr. Gabrielle Ronnett, Dr. Neil Copeland, and Dr. Nancy Jenkins for stimulating discussions. This work was supported by grants from the U.S. Public Health Service (AG 05146, NS 20471), The Adler Foundation, The Alzheimer's Association and the American Health Assistance Foundation.

References

Anderson JP, Esch FS, Keim PS, Sambamurti K, Lieberburg I, Robakis NK (1991) Exact cleavage site of Alzheimer amyloid precursor in neuronal PC-12 cells. Neurosci Lett 128:126–128

Borchelt DR, Shen J, Johannsdottir R, Kitt CA, Mojekwu JC, Thinakaran G, Sisodia SS, Price DL, Carlson G, Hsiao KK (1994) Premature death in transgenic mice expressing the Alzheimer's amyloid precursor protein. Soc Neurosci Abstr 20:636

Busciglio J, Gabuzda DH, Matsudaira P, Yankner BA (1993) Generation of β-amyloid in the secretory pathway in neuronal and nonneuronal cells. Proc Natl Acad Sci USA 90:2092–2096

Cai X-D, Golde TE, Younkin SG (1993) Release of excess amyloid β protein from a mutant amyloid β protein precursor. Science 259:514–516

Chartier-Harlin M-C, Crawford F, Houlden H, Warren A, Hughes D, Fidani L, Goate A, Rossor M, Roques P, Hardy J, Mullan M (1991) Early-onset Alzheimer's disease caused by mutations at codon 717 of the β-amyloid precursor protein gene. Nature 353:844–846

Citron M, Oltersdorf T, Haass C, McConlogue L, Hung AY, Seubert P, Vigo-Pelfrey C, Lieberburg I, Selkoe DJ (1992) Mutation of the β-amyloid precursor protein in familial Alzheimer's disease increases β-protein production. Nature 360:672–674

Cole GM, Huynh TV, Saitoh T (1989) Evidence for lysosomal processing of amyloid β-protein precursor in cultured cells. Neurochem Res 14:933–939

Cras P, Kawai M, Lowery D, Gonzalez-DeWhitt P, Greenberg B, Perry G (1991) Senile plaque neurites in Alzheimer disease accumulate amyloid precursor protein. Proc Natl Acad Sci USA 88:7552–7556

Esch FS, Keim PS, Beattie EC, Blacher RW, Culwell AR, Oltersdorf T, McClure D, Ward PJ (1990) Cleavage of amyloid β peptide during constitutive processing of its precursor. Science 248:1122–1124

Games D, Adams D, Alessandrini R, Barbour R, Berthelette P, Blackwell C, Carr T, Clemens J, Donaldson T, Gillespie F, Guido T, Hagopian S, Johnson-Wood K, Khan K, Lee M, Leibowitz P, Lieberburg I, Little S, Masliah E, McConlogue L, Montoya-Zavala M, Mucke L, Paganini L, Penniman E, Power M, Schenk D, Seubert P, Snyder B, Soriano F, Tan H, Vitale J, Wadsworth S, Wolozin B, Zhao J (1995) Alzheimer-type neuropathology in transgenic mice overexpressing V717F β-amyloid precursor protein. Nature 373:523–527

Glenner GG, Wong CW (1984) Alzheimer's disease: initial report of the purification and characterization of a novel cerebrovascular amyloid protein. Biochem Biophys Res Commun 120:885–890

Goate A, Chartier-Harlin M-C, Mullan M, Brown J, Crawford F, Fidani L, Giuffra L, Haynes A, Irving N, James L, Mant R, Newton P, Rooke K, Roques P, Talbot C, Pericak-Vance M, Roses A, Williamson R, Rossor M, Owen M, Hardy J (1991) Segregation of a missense mutation in the amyloid precursor protein gene with familial Alzheimer's disease. Nature 349:704–706

Golde TE, Estus S, Younkin LH, Selkoe DJ, Younkin SG (1992) Processing of the amyloid protein precursor to potentially amyloidogenic derivatives. Science 255:728–730

Haass C, Selkoe DJ (1993) Cellular processing of β-amyloid precursor and the genesis of amyloid β-peptide. Cell 75:1039–1042

Haass C, Hung AY, Selkoe DJ (1991) Processing of β-amyloid precursor protein in microglia and astrocytes favors an internal localization over constitutive secretion. J Neurosci 11:3783–3793

Haass C, Koo EH, Mellon A, Hung AY, Selkoe DJ (1992a) Targeting of cell-surface β-amyloid precursor protein to lysosomes: alternative processing into amyloid-bearing fragments. Nature 357:500–503

Haass C, Schlossmacher MG, Hung AY, Vigo-Pelfrey C, Mellon A, Ostaszewski BL, Lieberburg I, Koo EH, Schenk D, Teplow DB, Selkoe DJ (1992b) Amyloid β-peptide is produced by cultured cells during normal metabolism. Nature 359:322–325

Haass C, Hung AY, Selkoe DJ, Teplow DB (1994) Mutations associated with a locus for familial Alzheimer's disease result in alternative processing of amyloid β-protein precursor. J Biol Chem 269:17741–17748

Hendricks L, van Duijn CM, Cras P, Cruts M, Van Hul W, van Harskamp F, Warren A, McInnis MG, Antonarakis SE, Martin J-J, Hofman A, Van Broeckhoven C (1992) Presenile dementia and cerebal haemorrhage linked to a mutation at codon 692 of the β-amyloid precursor protein gene. Nature Genet 1:218–221

Higgins GA, Lewis DA, Bahmanyar S, Goldgaber D, Gajdusek DC, Young WG, Morrison JH, Wilson MC (1988) Differential regulation of amyloid-β-protein mRNA expression within hippocampal neuronal subpopulations in Alzheimer disease. Proc Natl Acad Sci USA 85:1297–1301

Higgins LS, Holtzman DM, Rabin J, Mobley WC, Cordell B (1994) Transgenic mouse brain histopathology resembles early Alzheimer's disease. Ann Neurol 35:598–607

Hung AY, Selkoe DJ (1994) Selective ectodomain phosphorylation and regulated cleavage of β-amyloid precursor protein. EMBO J 13:534–542

Iwatsubo T, Odaka A, Suzuki N, Mizusawa H, Nukina N, Ihara Y (1994) Visualization of Aβ42(43)-positive and Aβ40-positive senile plaques with end-specific Aβ-monoclonal antibodies: evidence that an initially deposited Aβ species is Aβ1–42(43). Neuron 13:45–53

Jarrett JT, Lansbury PT Jr (1993) Seeding "one-dimensional crystallization" of amyloid: a pathogenic mechanism in Alzheimer's disease and scrapie? Cell 73:1055–1058

Jarrett JT, Berger EP, Lansbury PT Jr (1993) The carboxy terminus of the β amyloid protein is critical for the seeding of amyloid formation: implications for the pathogenesis of Alzheimer's disease. Biochemistry 32:4693–4697

Kemper T (1984) Neuroanatomical and neuropathological changes in normal aging and in dementia. In: Albert ML (ed) Clinical neurology of aging. Oxford University Press, New York, pp 9–52

Klier FG, Cole G, Stalleup W, Schubert D (1990) Amyloid β-protein precursor is associated with extracellular matrix. Brain Res 515:336–342

Koo EH, Squazzo SL (1994) Evidence that production and release of amyloid β-protein involves the endocytic pathway. J Biol Chem 269:17386–17389

Koo EH, Sisodia SS, Archer DR, Martin LJ, Weidemann A, Beyreuther K, Fischer P, Masters CL, Price DL (1990a) Precursor of amyloid protein in Alzheimer disease undergoes fast anterograde axonal transport. Proc Natl Acad Sci USA 87:1561–1565

Koo EH, Sisodia SS, Cork LC, Unterbeck A, Bayney RM, Price DL (1990b) Differential expression of amyloid precursor protein mRNAs in case of Alzheimer's disease and in aged nonhuman primates. Neuron 2:97–104

LaFerla FM, Tinkle BT, Bieberich CJ, Haudenschild CC, Jay G (1995) The Alzheimer's Aβ peptide induces neurodegeneration and apoptotic cell death in transgenic mice. Nature Genet 9:21–30

Lamb BT, Sisodia SS, Lawler AM, Slunt HH, Kitt CA, Kearns WG, Pearson PL, Price DL, Gearhart JD (1993) Introduction and expression of the 400 kilobase precursor amyloid protein gene in transgenic mice. Nature Genet 5:22–30

Levy-Lahad E, Wijsman EM, Nemens E, Anderson L, Goddard KAB, Weber JL, Bird TD, Schellenberg GD (1995) A familial Alzheimer's disease locus on chromosome 1. Science 269:970–973

Martin LJ, Sisodia SS, Koo EH, Cork LC, Dellovade TL, Weidemann A, Beyreuther K, Masters C, Price DL (1991) Amyloid precursor protein in aged nonhuman primates. Proc Natl Acad Sci USA 88:1461–1465

Masters CL, Multhaup G, Simms G, Pottgiesser J, Martins RN, Beyreuther K (1985a) Neuronal origin of a cerebral amyloid: neurofibrillary tangles of Alzheimer's disease contain the same protein as the amyloid of plaque cores and blood vessels. EMBO J 4:2757–2763

Masters CL, Simms G, Weinman NA, Multhaup G, McDonald BL, Beyreuther K (1985b) Amyloid plaque core protein in Alzheimer disease and Down syndrome. Proc Natl Acad Sci USA 82:4245–4249

Mattson MP, Barger SW, Cheng B, Lieberburg I, Smith-Swintosky VL, Rydel RE (1993a) β-Amyloid precursor protein metabolites and loss of neuronal Ca^{2+} homeostasis in Alzheimer's disease. Trends Neurosci 16:409–414

Mattson MP, Cheng B, Culwell AR, Esch FS, Lieberburg I, Rydel RE (1993b) Evidence for excitoprotective and intraneuronal calcium-regulating roles for secreted forms of the β-amyloid precursor protein. Neuron 10:243–254

Moran PM, Higgins LS, Cordell B, Moser PC (1995) Age-related learning deficits in transgenic mice expressing the 751-amino acid isoform of human β-amyloid precursor protein. Proc Natl Acad Sci USA 92:5341–5345

Mucke L, Masliah E, Johnson WB, Ruppe MD, Alford M, Rockenstein EM, Forss-Petter S, Pietropaolo M, Mallory M, Abraham CR (1994) Synapto-trophic effects of human amyloid β protein precursors in the cortex of transgenic mice. Brain Res 666:151–167

Mucke L, Abraham CR, Ruppe MD, Rockenstein EM, Toggas SM, Mallory M, Alford M, Masliah E (1995) Protection against HIV-1 gp120-induced brain damage by neuronal expression of human amyloid precursor protein (hAPP). J Exp Med 181:1551–1556

Mullan M, Crawford F, Axelman K, Houlden H, Lillius L, Winblad B, Lann-felt L (1992) A pathogenic mutation for probable Alzheimer's disease in the APP gene at the N-terminus of β-amyloid. Nature Genet 1:345–347

Naruse S, Igarashi S, Kobayashi H, Aoki K, Inuzuka T, Kaneko K, Shimizu T, Iihara K, Kojima T, Miyatake T, Tsuji S (1991) Mis-sense mutation Val-Ile in exon 17 of amyloid precursor protein gene in Japanese familial Alz-heimer's disease. Lancet 337:978–979

Neve RL, Finch EA, Dawes LR (1988) Expression of the Alzheimer amyloid precursor gene transcripts in the human brain. Neuron 1:669–677

Oltersdorf T, Fritz LC, Schenk DB, Lieberburg I, Johnson-Wood KL, Beattie EC, Ward PJ, Blacher RW, Dovey HF, Sinha S (1989) The secreted form of the Alzheimer's amyloid precursor protein with the Kunitz domain is protease nexin-II. Nature 341:144–147

Palmert MR, Golde TE, Cohen ML, Kovacs DM, Tanzi RE, Gusella JF, Usiak MF, Younkin LH, Younkin SG (1988) Amyloid protein precursor mess-enger RNAs: differential expression in Alzheimer's disease. Science 241:1080–1084

Price DL, Sisodia SS (1994) Cellular and molecular biology of Alzheimer's disease and animal models. Annu Rev Med 45:435–446

Probst A, Langui D, Ipsen S, Robakis N, Ulrich J (1991) Deposition of β/A4 protein along neuronal plasma membranes in diffuse senile plaques. Acta Neuropathol (Berl) 83:21–29

Rogaev EI, Sherrington R, Rogaeva EA, Levesque G, Ikeda M, Liang Y, Chi H, Lin C, Holman K, Tsuda T, Mar L, Sorbi S, Nacmias B, Piacentini S, Amaducci L, Chumakov I, Cohen D, Lannfelt L, Fraser PE, Rommens JM,

St George-Hyslop PH (1995) Familial Alzheimer's disease in kindreds with missense mutations in a gene on chromosome 1 related to the Alzheimer's disease type 3 gene. Nature 376:775–778

Saitoh T, Sundsmo M, Roch J-M, Kimura T, Cole G, Schubert D, Oltersdorf T, Schenk DB (1989) Secreted form of amyloid β protein precursor is involved in the growth regulation of fibroblasts. Cell 58:615–622

Sandbrink R, Masters CL, Beyreuther K (1994) Similar alternative splicing of a non-homologous domain in βA4-amyloid protein precursor-like proteins. J Biol Chem 269:14227–14234

Schubert D, Jin L-W, Saitoh T, Cole G (1989) The regulation of amyloid β protein precursor secretion and its modulatory role in cell adhesion. Neuron 3:689–694

Selkoe DJ, Bell DS, Podlisny MB, Price DL, Cork LC (1987) Conservation of brain amyloid proteins in aged mammals and humans with Alzheimer's disease. Science 235:873–877

Seubert P, Oltersdorf T, Lee MG, Barbour R, Blomquist C, Davis DL, Bryant K, Fritz LC, Galasko D, Thal LJ, Lieberburg I, Schenk DB (1993) Secretion of β-amyloid precursor protein cleaved at the amino terminus of the β-amyloid peptide. Nature 361:260–263

Sherrington R, Rogaev EI, Liang Y, Rogaeva EA, Levesque G, Ikeda M, Chi H, Lin C, Li G, Holman K, Tsuda T, Mar L, Foncin J-F, Bruni AC, Montesi MP, Sorbi S, Rainero I, Pinessi L, Nee L, Chumakov I, Pollen D, Brookes A, Sanseau P, Polinsky RJ, Wasco W, Da Silva HAR, Haines JL, Pericak-Vance MA, Tanzi RE, Roses AD, Fraser PE, Rommens JM, St George-Hyslop PH (1995) Cloning of a gene bearing missense mutations in early-onset familial Alzheimer's disease. Nature 375:754–760

Shivers BD, Hilbich C, Multhaup G, Salbaum M, Beyreuther K, Seeburg PH (1988) Alzheimer's disease amyloidogenic glycoprotein: expression pattern in rat brain suggests a role in cell contact. EMBO J 7:1365–1370

Shoji M, Golde TE, Ghiso J, Cheung TT, Estus S, Shaffer LM, Cai X-D, McKay DM, Tintner R, Frangione B, Younkin SG (1992) Production of the Alzheimer amyloid β protein by normal proteolytic processing. Science 258:126–129

Sisodia SS (1992) β-amyloid precursor protein cleavage by a membrane-bound protease. Proc Natl Acad Sci USA 89:6075–6079

Sisodia SS, Koo EH, Beyreuther K, Unterbeck A, Price DL (1990) Evidence that β-amyloid protein in Alzheimer's disease is not derived by normal processing. Science 248:492–495

Sisodia SS, Koo EH, Hoffman PN, Perry G, Price DL (1993) Identification and transport of full-length amyloid precursor proteins in rat peripheral nervous system. J Neurosci 13:3136–3142

Slunt HH, Thinakaran G, von Koch C, Lo ACY, Tanzi RE, Sisodia SS (1994) Expression of a ubiquitous, cross-reactive homologue of the mouse β-amyloid precursor protein (APP). J Biol Chem 269:2637–2644

Suzuki N, Cheung TT, Cai X-D, Odaka A, Otvos L Jr, Eckman C, Golde TE, Younkin SG (1994) An increased percentage of long amyloid β protein secreted by familial amyloid β protein precursor (βAPP$_{717}$) mutants. Science 264:1336–1340

Thinakaran G, Sisodia SS (1994) Amyloid precursor-like protein 2 (APLP2) is modified by the addition of chondroitin sulfate glycosaminoglycan at a single site. J Biol Chem 269:22099–22104

Thinakaran G, Slunt HH, Sisodia SS (1995) Novel regulation of chondroitin sulfate glycosaminoglycan modification of amyloid precursor protein and its homologue, APLP2. J Biol Chem 270:16522–16525

Thinakaran G, Roskams AJI, Kitt CA, Slunt HH, Masliah E, von Koch C, Ginsberg S, Ronnett GV, Reed RR, Price DL, Sisodia SS (1996) Distribution of an APP homologue, APLP2, in the mouse olfactory system; a potential role for APLP2 in axogenesis. J Neurosci 15:6314–6326

Wang R, Meschia JF, Cotter RJ, Sisodia SS (1991) Secretion of the β/A4 amyloid precursor protein. Identification of a cleavage site in cultured mammalian cells. J Biol Chem 266:16960–16964

Wasco W, Bupp K, Magendantz M, Gusella JF, Tanzi RE, Solomon F (1992) Identification of a mouse brain cDNA that encodes a protein related to the Alzheimer disease-associated amyloid-beta-protein precursor. Proc Natl Acad Sci USA 89:10758–10762

Wasco W, Gurubhagavatula S, Paradis Md, Romano DM, Sisodia SS, Hyman BT, Neve RL, Tanzi RE (1993) Isolation and characterization of APLP2 encoding a homologue of the Alzheimer's associated amyloid β protein precursor. Nature Genet 5:95–99

Weidemann A, König G, Bunke D, Fischer P, Salbaum JM, Masters CL, Beyreuther K (1989) Identification, biogenesis, and localization of precursors of Alzheimer's disease A4 amyloid protein. Cell 57:115–126

Wisniewski HM, Terry RD (1973) Morphology of the aging brain, human and animal. Prog Brain Res 40:167–186

Zheng H, Jiang M-H, Trumbauer ME, Sirinathsinghji DJS, Hopkins R, Smith DW, Heavens RP, Dawson GR, Boyce S, Conner MW, Stevens KA, Slunt HH, Sisodia SS, Chen HY, Van der Ploeg LHT (1995) β-amyloid precursor protein-deficient mice show reactive gliosis and decreased locomotor activity. Cell 81:525–531

5 Molecular Processing Pathways of β-Amyloid Precursor Protein: Therapeutic Implications

C. Haass

5.1 The Pathobiology of Alzheimer's Disease

Alzheimer's disease (AD) is the most commonly occurring neurode-generative disorder. In most of the cases, AD occurs sporadically; however, in a small percentage of AD (familial AD; FAD) mutations were found to cosegregate with early onset of AD.

Pathologically, AD is characterized by the accumulation of numerous extracellular plaques in the brain parenchyma and depositions in meningeal and cortical blood vessels (for review, see Selkoe

1994a,b). Senile plaques are surrounded by degenerating neurons, indicating that the plaque components have a toxic effect on surrounding nerve cells. The major component of senile plaques is the 4-kDa amyloid β-peptide (Aβ). Aβ is a 39–42 amino acid peptide; however, most of the peptides begin with aspartate 1 and end with amino acid 40 (reviewed by Selkoe 1994a,b). The peptide is very hydrophobic and spontaneously forms insoluble aggregates. Aβ was shown to be neurotoxic and selectively kills cultured neurons derived from mouse or human brains (Yankner et al. 1990). Moreover, Aβ-neurotoxicity depends on the ability of Aβ to form insoluble aggregates. It was shown that only these aggregates exhibit neurotoxicity, whereas the soluble monomer is not toxic (Lorenzo and Yankner 1994).

Cloning of the corresponding gene from which Aβ is derived showed that Aβ is proteolytically generated from a high-molecular-weight precursor protein, the β-amyloid precursor protein (βAPP; Kang et al. 1987).

5.2 Amyloid β-Peptide Is Derived from a Precursor Protein by Proteolytic Processing

βAPP is a type I transmembrane protein with a single transmembrane domain (Fig. 1a; Dyrks et al. 1988), a small cytoplasmic C-terminal domain, and large extracellular N-terminal domain. Aβ itself is only a very small portion of this molecule, indicating that it is generated by proteolytic processing of its precursor (Kang et al. 1987). However, this is complicated by the fact that the C-terminus of Aβ is embedded in the transmembrane domain, protecting it from proteolytic cleavage. Moreover, there is a major processing pathway of βAPP which clearly inhibits Aβ generation. βAPP is proteolytically cleaved by an enzyme called α-secretase leading to the secretion of the N-terminal portion of the precursor, designated soluble APP (APP$_s$; Fig. 1b). APP$_s$ is then found in conditioned media of cultured cells or in vivo within the cerebrospinal fluid and plasma (Weidemann et al. 1989; Sisodia et al. 1990). α-Secretase cleavage predominantly occurs at position 16/17 of Aβ, thus clearly inhibiting Aβ production (Esch et al. 1990). A 10-kDa C-terminal fragment of βAPP remains within the membrane (Fig. 1b). Based on these observations, it was argued that only aberrant processing

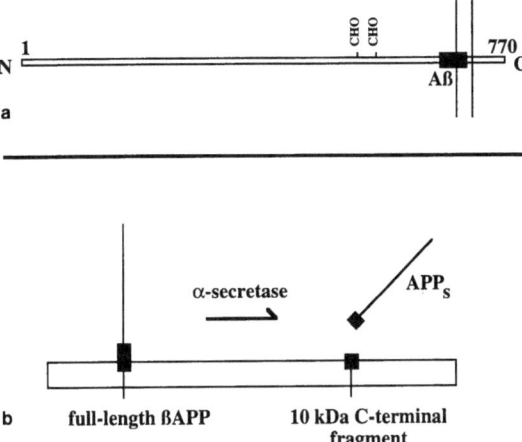

Fig. 1a,b. Schematic representation of the β-amyloid precursor protein (βAPP) molecule. **a** The *black box* represents the amyloid β-peptide (Aβ) domain. The *vertical lines* indicate the plasma membrane. **b** Processing of βAPP by α-secretase leads to the secretion of soluble APP (*APPs*) and the generation of a 10-kDa C-terminal fragment. *N*, N-terminus; *C*, C-terminus

of βAPP under pathological conditions could lead to the formation of Aβ.

5.3 The Endosomal/Lysosomal Processing Pathway of βAPP

In contrast to the proposed hypothesis that only pathological processing of βAPP could lead to the production of Aβ, βAPP degradation products containing the entire Aβ domain were found within the brains of AD patients and healthy controls (Estus et al. 1992; Tamaoka et al. 1992) and within cultured cells transfected with the βAPP gene (Golde et al. 1991; Haass et al. 1992a). In cell culture, these fragments where found to be stabilized and to accumulate upon treatment with cystein protease inhibitors such as leupeptin or E64 (Golde et al. 1991; Haass et al. 1992a). Since endosomes and lysosomes are particulary enriched in cystein proteases, lysosomes were isolated from βAPP transfected cells and shown to contain a high level of amyloidogenic βAPP fragments

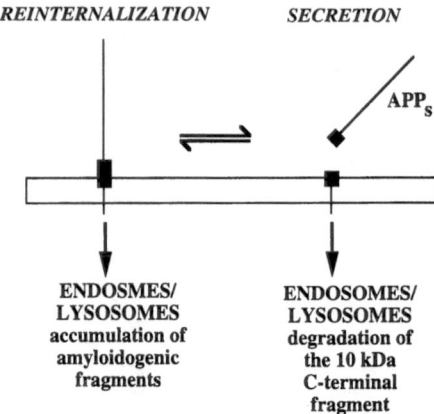

Fig. 2. Full-length β-amyloid precursor protein (*left*) or the 10-kDa C-terminal fragment (*right*) can be reinternalized on the cell surface and targeted to endosomes and lysosomes. *APP_s,* soluble amyloid precursor protein

between 12 and 28 kDa (Haass et al. 1992a). These fragments contain the cytoplasmic tail of βAPP and the entire Aβ domain, thus representing N-terminally degraded βAPP fragments. Some of these fragments were radiosequenced, and a 12-kDa fragment was shown to have an N-terminus beginning at Asp 1 (Cheung et al. 1994), which is the N-terminus of Aβ isolated from senile plaques. This amyloidogenic precursor of Aβ generation can clearly be formed in vivo and in cultured cells under physiological conditions, therefore it was speculated in contrast to the previous hypothesis that Aβ generation might also occur under normal circumstances.

How is βAPP targeted to endosomes/lysosomes? βAPP is reinternalized (Fig. 2; Haass et al. 1992a) from the cell surface into clathrin-coated pits (Nordstedt et al. 1993) and from there to endosomes and lysosomes. Reinternalization seems to be dependent on a dual tyrosine-containing sequence motif within the cytoplasmic tail of βAPP (Lai et al. 1995; Haass and Capell, unpublished data). Therefore, reinternalization targets βAPP away from α-secretase activity, making full-length βAPP accessible for potential Aβ generation.

5.4 Physiologic Production of Amyloid β-Peptide

Indeed, Aβ has been shown to be secreted by a physiological pathway in numerous cultured human cell lines tested so far (Haass et al. 1992b; Shoji et al. 1992; Busciglio et al. 1993). Stably transfected cell lines were metabolically labeled with ^{35}S-methionine for up to 12–24 h. Conditioned media were then immunoprecipitated with antibodies to synthetic Aβ. This clearly identified a 4-kDa peptide, which subsequently was proven to be authentic Aβ and a 3-kDa peptide, designated p3. Radiosequencing of the 4-kDa peptide (Fig. 3) showed that the majority of these peptides begin with an N-terminus at Asp 1 of Aβ (Haass et al. 1992b; Shoji et al. 1992) and predominantly terminate at amino acid 40 (Suzuki et al. 1994) which is identical to Aβ isolated from senile plaques. However, some heterogeneity was observed at both ends of the molecule as indicated by the presence of N-terminally elongated/truncated peptides or Aβ peptides ending at amino acid 42 instead of amino acid 40. While the heterogeneity seemed to be of minor

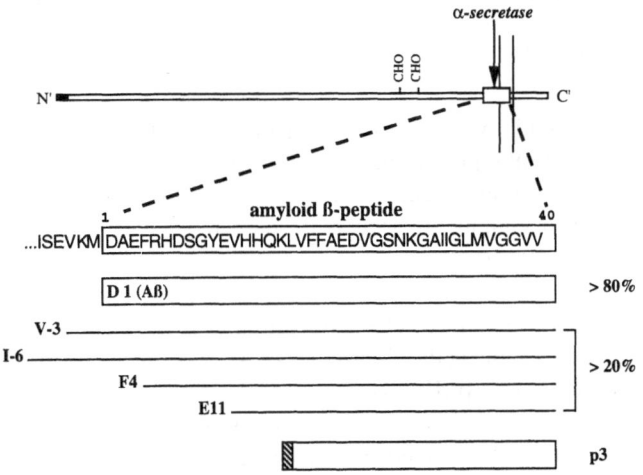

Fig. 3. Radiosequencing of amyloid β-peptide (Aβ) and p3. The majority of Aβ peptides produced by cultured cells have an N-terminus at Asp 1 (>80%) similar to Aβ isolated from senile plaques. Small amounts of peptides elongated or truncated at the N-terminus are also found. p3 has a N-terminus at Leu 17 or Val 18. *C*, C-terminus; *CHO*, glycosylation sites

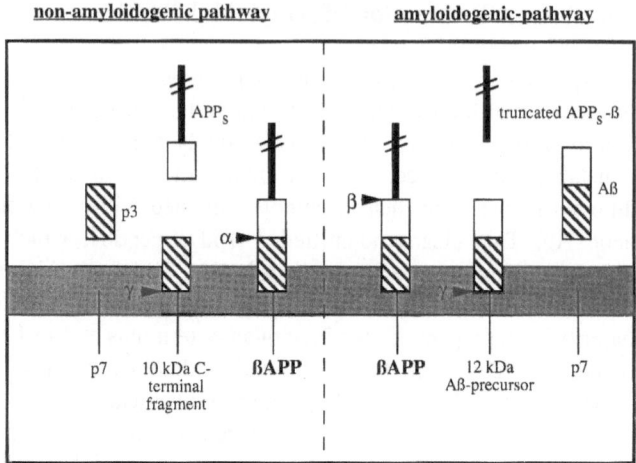

non-amyloidogenic pathway   amyloidogenic-pathway

Fig. 4. Proteolytic processing of β-amyloid precursor protein (βAPP) can be divided into two different pathways, an amyloidogenic (*right panel*) and a nonamyloid~genic pathway (*left panel*). The C-terminal portion of the amyloid β-peptide (Aβ) domain giving rise to p3 is shown as a *hatched box*. p7 is a putative proteolytic product of the γ-secretase cleavage. *APP$_s$,* soluble amyloid precursor protein

importance initially, it is now clear that peptides elongated specifically at the C-terminus play a major role in the pathogenesis of AD (see below). Interestingly, radiosequencing of p3 revealed that it has an N-terminus beginning precisely at the α-secretase cleavage site (Haass et al. 1992b, 1993).

The finding that Aβ is secreted physiologically suggests that there may be two additional secretases (reviewed by Haass and Selkoe 1993): β-secretase cleaving at the N-terminus of Aβ and γ-secretase cleaving at the C-terminus of Aβ (Fig. 4). The proteolytic processing of βAPP can now be divided into two separate pathways, a nonamyloidogenic pathway and an amyloidogenic pathway (Fig. 4). In the latter pathway, βAPP is cleaved by β-secretase at the N-terminus of the Aβ domain,which results in the generation of a 12-Da amyloidogenic fragment and the formation of a C-terminal truncated form of APP$_s$, called APP$_s$-β (Seubert et al. 1993). Such amyloidogenic precursors accumulate within endosomes and lysosomes (Haass et al. 1992a). The 12-kDa precursor is then cleaved by an unknown mechanism by γ-secretase, liberating Aβ into the culture media.

In the nonamyloidogenic pathway, βAPP is cleaved by α-secretase, whereby APP$_s$ is secreted and a 10-kDa C-terminal fragment generated as described above. The 10-kDa C-terminal fragment, the precursor for p3 generation (Haass et al. 1993), is cleaved by γ-secretase, similar to the way amyloidogenic 12-kDa precursor is cleaved, which then leads to the secretion of p3 into the media (Fig. 4).

The tissue culture system therefore provides an optimal model system for analyzing the molecular mechanism of Aβ generation and the influence of βAPP mutations on Aβ production. Moreover, components lowering Aβ production can be screened using the simple tissue culture system. The importance of the tissue culture system is supported by the finding that soluble Aβ is not only found within the conditioned media of cultured cells, but also in biological fluids such as cerebrospinal fluid and plasma (Seubert et al. 1992; Shoji et al. 1992).

5.5 The Molecular Mechanism
of Amyloid β-Peptide Production

To determine whether reinternalization of full-length βAPP is necessary for Aβ production, the cytoplasmic tail of βAPP was deleted, which results in an inhibition of βAPP endocytosis (Haass et al. 1993; Koo and Squazzo 1994). Under these conditions, we found that kidney 293 cells still produced high levels of Aβ. However, radiosequencing the N-termini of these peptides revealed a strong inhibition of the authentic cleavage at Asp 1. This is paralleled by a marked increase of alternatively cleaved Aβ peptides which are truncated or elongated at the N-terminus (Haass et al. 1993), indicating that reinternalization is necessary for the correct N-terminal cleavage of Aβ at Asp 1. Moreover, in other cell types, such as Chinese hamster ovary (CHO) cells, which show a reduced ability of alternative Aβ cleavage, an overall inhibition of Aβ was observed upon inhibition of βAPP reinternalization (Koo and Squazzo 1994). This leads to the following model of Aβ generation: βAPP is reinternalized from the cell surface and targeted to early endosomes. The β-secretase cleavage occurs within early endosomes, creating a 12-kDa amyloidogenic precursor and a truncated form of APP$_s$, called APP_s-β. Some of the endosomes recycle back to the cell surface, resulting in the secretion of APP_s-β (Seubert et al. 1993) and the appearance of the 12-kDa precursor on the cell-surface (Koo and Squazzo 1994). The precursor is then cleaved by γ-secretase by an unknown mechanism which results in the ultimate secretion of $A\beta$ into the culture media. However, as described below, $A\beta$ can also be generated by a different cellular mechanism, at least in the case of one of the FAD mutations.

5.6 Amyloid β-Peptide Production Is Inhibited by Activation of Protein Kinase C

Activation of protein kinase C (PKC) results in an increased secretion of APP_s (Caporaso et al. 1992; Nitsch et al. 1992), suggesting that the α-secretory pathway is activated. One could therefore argue that increased secretion of APP_s might inhibit the amyloidogenic processing of βAPP and therefore the generation of $A\beta$. When kidney 293 cells stably transfected with the βAPP cDNA are treated with phorbolesters one indeed finds a strong inhibition in $A\beta$ production by more than 60% (Hung et al. 1993; Fig. 5a). In parallel, secretion of p3 is strongly enhanced (Hung et al. 1993). This is expected, because activation of the α-secretory pathway results in an enhanced production of the 10-kDa C-terminal fragment, which is the membrane-bound precursor for p3 (Figs. 4, 5b; Haass et al. 1993). A similar result is observed when one stimulates muscarinic m1 and m3 acetylcholine receptors transfected into βAPP-expressing cells with carbachol (Hung et al. 1993). Therefore, pharmacological activation of first messenger systems coupled to PKC might prove useful in lowering $A\beta$ production, which can now be tested in animal models. However, it should be noted that activation of PKC results in a strong increase of the very hydrophobic p3. Moreover, nothing is known about neurotoxic properties of p3 or about its influence on $A\beta$ aggregation. Therefore, any potential therapeutic approaches using PKC activation should be undertaken with care.

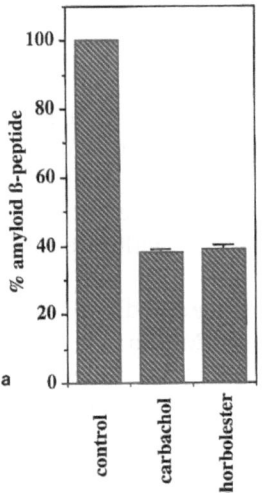

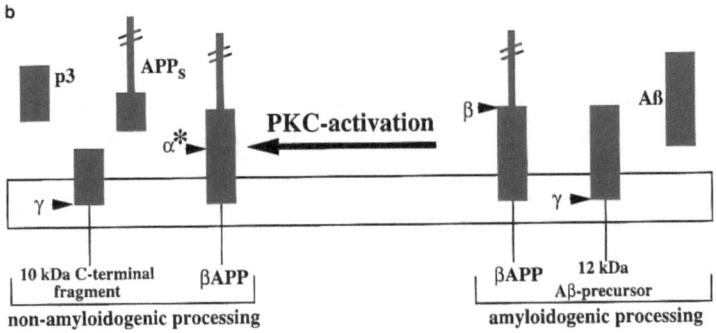

Fig. 5a,b. Stimulation of protein kinase C (*PKC*) with phorbolesters or stimulation of the muscarinic receptor m1 with carbachol inhibit amyloid β-peptide (Aβ) production by approximately 60% (**a**) and results in an enhanced secretion of p3 and APP$_s$ (**b**). α*, activated α-secretory pathway. βAPP, β-amyloid precursor protein; *APP$_s$*, soluble APP

5.7 Mutations Within the βAPP Gene Influence Proteolytic Processing of Amyloid β-Peptide

As mentioned above, early onset FAD is inherited as an autosomal dominant condition. In some rare cases of FAD, mutations were found within the βAPP gene (for review see Mullan and Crawford 1993). These mutations occur close to or at the three secretase cleavage sites described above (Fig. 6) and directly influence Aβ generation (Table 1).

1. A double mutation, called the Swedish mutation, occurs at the N-terminus of the Aβ domain (Mullan et al. 1992). The mutation results in a 3- to 6-fold increase of Aβ production in transfected cells (Citron et al. 1992; Cai et al. 1993) and in primary fibroblasts cultured from members of the Swedish family (Citron et al. 1994). Cell biologically, it was shown that this mutation results in earlier β-secretase cleavage as in the case of the wild-type βAPP gene. An antibody specifically recognizing the C-terminus of the truncated APP$_s$-β^{sw} was used to monitor the timing and cellular localization of the β-secretase cleavage of Swedish βAPP (Haass et al. 1995). Using this antibody, the β-secretase cleavage of Swedish βAPP was shown to occur predominantly within Golgi-derived vesicles, most likely secretory vesicles. In contrast, β-secretase cleavage of wild-

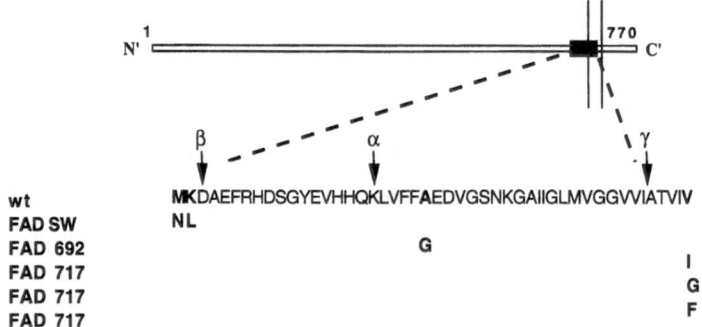

Fig. 6. Familial Alzheimer's disease (*FAD*) mutations within the β-amyloid precursor protein (βAPP) gene are found within or around the amyloid β-peptide (Aβ) domain close to the three secretase cleavage sites. *N*, N-terminus; *C*, C-terminus; *NL, G, A, F,* FAD mutations

Table 1. The effect of FAD mutations within the βAPP gene on Aβ generation

FAD gene mutation	Influence on Aβ generation	Molecular mechanism	Pathogenic mechanism
MK/NL	Three- to sixfold increase of Aβ production	Mutation provodes a better substrate for β-secretase, resulting in an early cleavage within secretory vesicles	Overproduction of Aβ leads to earlier plaque formation
A/G	Modest increase of Aβ production and generation of N-terminal, truncated Aβ-like peptides	Inhibition of α-scretase activity	N-terminal, truncated peptides might aggregate faster, which leads to earlier plaque formation
V/I V/F V/G	Increased amounts of Aβ peptides ending at amino acid 42 instead of 40	Mutations might provide a new recognition sequence for γ-secretase	Aβ peptides ending at amino acid 52 aggregate faster, which leads to earlier plaque formation

FAD, familial Alzheimer's disease; *Aβ*, amyloid β-peptide.

type βAPP occurs within a later compartment, the early endosomes. The earlier cleavage seems to be due to better substrate created by the mutation at the β-secretase cleavage site (Citron et al. 1995). Thus, in the case of the Swedish mutation early onset of AD seems to be explained by the strong increase in Aβ production, which leads to early amyloid deposition.

2. A second mutation is localized close to the α-secretase cleavage site (Hendriks et al. 1992). This mutation results in increased production of Aβ and Aβ-like peptides with alternative N-termini (Haass et al. 1994). Such truncated Aβ-like peptides might aggregate more rapidly (Jarret and Lansbury 1993). These truncated peptides might therefore initiate and accelerate amyloid plaque formation.

3. Three mutations are found at the C-terminus of the Aβ domain (Goate et al. 1991; Chartier-Harlin et al. 1991; Murrel et al. 1991). These mutations cause cleavages at the Aβ C-terminus that result in an enhanced production of Aβ peptides ending at position 42 instead of position 40 (Suzuki et al. 1994). Elongated Aβ peptides are known to aggregate more rapidly and might thus accelerate amyloid plaque formation (Jarret and Lansbury 1993).

Taken together, mutations within the βAPP gene increase Aβ production itself or lead to the production of Aβ peptides elongated at the C-terminus or truncated at the N-terminus. In either case amyloid plaque formation is induced at very early time points due to enhanced Aβ production and/or Aβ aggregation (Table 1).

5.8 Therapeutic Implications

Based on the data presented above, the following potential therapeutic strategies to prevent or slow down AD can be proposed. The primary goal is to find specific protease inhibitors for either β- or γ-secretase. Inhibition of γ-secretase is favorable, since that would inhibit not only the generation of Aβ, but also that of p3. A more indirect approach to inhibit Aβ generation would be to divert the amyloidogenic processing pathway of βAPP into the nonamyloidogenic pathway. This is possible by activation of PKC, as described above. However, this results in increased production of p3, which might be neurotoxic as well or might even stimulate Aβ aggregation.

A completely different approach would be to inhibit Aβ aggregation, which prevents the neurotoxic activity of Aβ. Such a strategy has been used by Lorenzo and Yankner (1994) who treated neurotoxic Aβ aggregates with Congo red. This treatment clearly prevented Aβ neurotoxicity; however, it is not clear whether Congo red treatment indeed inhibits Aβ aggregation.

An almost indefinite number of potential components can now be screened for their corresponding effects (inhibiting Aβ production or neurotoxicity of Aβ) in tissue culture systems such as the ones described above, and positive candidates used in animal models such as the transgenic mice model for AD described recently by Games et al. (1995).

References

Busciglio J, Gabuzda DH, Matsudaira P, Yankner BA (1993) Generation of β-amyloid in the secretory pathway in neuronal and nonneuronal cells. Proc Natl Acad Sci USA 90:2092–2096

Cai XD, Golde TE, Younkin SG (1993) Release of excess amyloid β protein from a mutant amyloid β protein precursor. Science 259:514–516

Caporaso GL, Gandy S, Buxbaum JD, Ramabhadran TV, Greengard P (1992) Protein phosphorylation regulates secretion of Alzheimer β/A4 amyloid precursor protein. Proc Natl Acad Sci USA 89:3055–3059

Chartier-Harlin MC, Crawford F, Houlden H, Warren A, Hughes D, Fidani L, Goate A, Rossor M, Roques P, Hardy J, Mullan M (1991) Early-onset Alzheimer's disease caused by mutations at codon 717 of the β-amyloid precursor protein gene. Nature 353:844–846

Cheung TT, Ghiso J, Shoji M, Cai XD, Golde T, Gandy SE, Frangione B, Younkin S (1994) Characterization by radiosequencing of the carboxyl-terminal derivatives produced from normal and mutant amyloid β protein precursors. Amyloid. Int J Exp Clin Invest 1:30–38

Citron M, Oltersdorf T, Haass C, McConlogue L, Hung AY, Seubert P, Vigo-Pelfrey C, Lieberburg I, Selkoe DJ (1992) Mutation of the β-amyloid precursor protein in familial Alzheimer's disease increases β-protein production. Nature 360:672–674

Citron M, Vigo-Pelfrey C, Teplow DB, Miller C, Schenk D, Johnston J, Winblad B, Venizelos N, Lannfelt L, Selkoe DJ (1994) Excessive production of amyloid β-protein by peripheral cells of symptomatic and presymptomatic patients carrying the Swedish familial Alzheimer disease mutation. Proc Natl Acad Sci USA 91:11993–11997

Citron M, Teplow DB, Selkoe DJ (1995) Generation of amyloid β protein from its precursor is sequence specific. Neuron 14:661–670

Dyrks T, Weidemann A, Multhaup G, Salbaum JM, Lemaire HG, Kang J, Müller-Hill B, Masters CL, Beyreuther K (1988) Identification, transmembrane orientation and biogenesis of the amyloid A4 precursor of Alzheimer's disease. EMBO J 7:949–957

Esch FS, Keim PS, Beattle EC, Blacher RW, Culwell AR, Oltersdorf T, McClure D, Ward PJ (1990) Cleavage of amyloid β peptide during constitutive processing of its precursor. Science 248:1122–1124

Estus S, Golde TE, Kunishita T, Blades D, Lowery D, Eisen M, Usiak M, Qu X, Tabira T, Greenberg BD, Younkin SG (1992) Potentially amyloidogenic, carboxyl-terminal derivatives of the amyloid protein precursor. Science 255:726–730

Games D, Adams D, Alessandrini R, Barbour R, Berthelette P, Blackwell C, Carr T, Clemens J, Donaldson T, Gillespie F, Guido T, Hagopain S, Johnson-Wood K, Khan K, Lee M, Leibowitz P, Lieberburg I, Little S, Masliah E, McConlogue L, Montoya-Zavala M, Mucke L, Paganini L, Penniman E, Power M, Schenk D, Seubert P, Snyder B, Soriano F, Tan H, Vitale J, Wadsworth S, Wolozin B, Zhao J (1995) Alzheimer type neuropathology in transgenic mice overexpressing V717F β-amyloid precursor protein. Nature 373:523–528

Goate A, Chartier-Harlin MC, Mullan M, Brown J, Crawford F, Fidani L, Giuffra L, Haynes A, Irving N, James L, Mant R, Newton P, Rooke K, Roques P, Talbot C, Pericak-Vance M, Roses A, Williamson R, Rossor M, Owen M, Hardy J (1991) Segregation of a missense mutation in the amyloid precursor gene with familial Alzheimer's disease. Nature 349:704–706

Golde TE, Estus S, Younkin LH, Selkoe DJ, Younkin S (1991) Processing of the amyloid protein precursor to potentially amyloidogenic derivatives. Science 255:728–730

Haass C, Selkoe DJ (1993) Cellular processing of β-amyloid precursor protein and the genesis of amyloid β-peptide. Cell 75:1039–1042

Haass C, Koo E, Mellon A, Hung AY, Selkoe D J (1992a) Targeting of cell-surface β-amyloid precursor protein to lysosomes: alternative processing into amyloid-bearing fragments. Nature 357:500–503

Haass C, Schlossmacher MG, Hung AY, Vigo-Pelfrey C, Mellon A, Ostaszewski BL, Lieberburg I, Koo EH, Schenk D, Teplow DB, Selkoe DJ (1992b) Amyloid β-peptide is produced by cultured cells during normal metabolism. Nature 359:322–325

Haass C, Hung, AY, Schlossmacher MG, Teplow DB, Selkoe DJ (1993) β-amyloid peptide and a 3 kDa fragment are derived by distinct cellular mechanisms. J Biol Chem 268:3021–3024

Haass C, Hung AY, Selkoe DJ, Teplow DB (1994) Mutations associated with a locus for familial Alzheimer's disease result in alternative processing of amyloid β-protein precursor. J Biol Chem 269:17741–17748

Haass C, Lemere CA, Capell A, Citron M, Seubert P, Schenk D, Lannfelt L, Selkoe DJ (1995) β-Secretase cleavage of β-amyloid precursor protein with the Swedish mutation occurs within the secretory pathway after the trans-Golgi network. Nature Med 1;1291–1296

Hendriks L, van Duijn CM, Cras P, Cruts M, van Hul W, van Harskamp F, Warren A, McInnis MG, Antonarakis SE, Martin JJ, Hofman A, van Brockhoeven C (1992) Presenile dementia and cerebral haemorrhage linked to a mutation at codon 692 of the β-amyloid precursor protein gene. Nature Genet 1:218–221

Hung AY, Haass C, Nitsch RM, Qiu WQ, Citron M, Wurtman RJ, Growdon JH, Selkoe DJ (1993) Activation of protein kinase C inhibits cellular production of the amyloid β-protein. J Biol Chem 268:22959–22962

Jarrett JT, Lansbury PT Jr (1993) Seeding "one-dimensional crystallization" of amyloid: a pathogenic mechanism in Alzheimer's disease and scrapie? Cell 73:1055–1058

Kang J, Lemaire HG, Unterbeck A, Salbaum MJ, Masters CL, Grzeschik K-H, Multhaup G, Beyreuther K, Müller-Hill B (1987) The precursor of Alzheimer's disease amyloid A4 protein resembles a cell-surface receptor. Nature 325:733–736

Koo EH, Squazzo SL (1994) Evidence that production and release of amyloid β-protein involves the endocytic pathway. J Biol Chem 269:17386–17389

Lai A, Sisodia SS, Trowbridge IS (1995) Characterization of sorting signals in the β-amyloid precursor protein cytoplasmic domain. J Biol Chem 270:3565–3573

Lorenzo A, Yankner BA (1994) β-Amyloid neurotoxicity requires fibril formation and is inhibited by congo red. Proc Natl Acad Sci USA 91:12243–12247

Mullan M, Crawford F (1993) Genetic and molecular advances in Alzheimer's disease. Trends Neurosci 16:398–414

Mullan M, Crawford F, Axelman K, Houlden H, Lilius L, Winblad B, Lannfelt L (1992) A pathogenic mutation for probable Alzheimer's disease in the APP gene at the N-terminus of β-amyloid. Nature Genet 1:345–347

Murrell J, Farlow M, Ghetti B, Benson MD (1991) A mutation in the amyloid precursor protein associated with hereditary Alzheimer's disease. Science 254:97–99

Nitsch RM, Slack BE, Wurtman RJ, Growdon JH (1992) Release of Alzheimer amyloid precursor derivatives stimulated by activation of muscarinic acetylcholine receptors. Science 258:304–307

Nordstedt C, Caporaso GL, Thyberg J, Gandy SE, Greengard P (1993) Identification of the Alzheimer β/A4 amyloid precursor protein in clathrin-coated
vesicles purified from PC 12 cells. J Biol Chem 268:608–612

Selkoe DJ (1994a) Normal and abnormal biology of the β-amyloid precursor
protein. Annu Rev Neurosci 17:489–517

Selkoe DJ (1994b) Cell biology of the amyloid β-protein precursor and the
mechanism of Alzheimer's disease. Annu Rev Cell Biol 10:373–403

Seubert P, Vigo-Pelfrey C, Esch F, Lee M, Dovey H, Davis D, Sinha S,
Schlossmacher M, Whaley J, Swindlehurst C, McCormack R, Wolfert R,
Selkoe DJ, Lieberburg I, Schenk D (1992) Isolation and quantification of
soluble Alzheimer's β-peptide from biological fluids. Nature 359:325–327

Seubert P, Oltersdorf T, Lee MG, Barbour R, Blomquist C, Davis DL, Bryant
K, Fritz LC, Galasko D, Thal LJ, Lieberburg I, Schenk DB (1993) Secretion of β-amyloid precursor protein cleaved at the amino terminus of the β-
amyloid peptide. Nature 361:260–263

Shoji M, Golde TE, Ghiso J, Cheung TT, Estus S, Shaffer LM, Cai XD,
McKay DM, Tintner R, Frangione B, Younkin SG (1992) Production of the
Alzheimer amyloid β protein by normal proteolytic processing. Science
258:126–129

Sisodia SS, Koo EH, Beyreuther K, Unterbeck A, Price DL (1990) Evidence
that β-amyloid protein in Alzheimer's disease is not derived by normal processing. Science 248:492–495

Suzuki N, Cheung TT, Cai XD, Odaka A, Otvos L jr, Eckman C, Golde TE,
Younkin SG (1994). An increased percentage of long amyloid β protein secreted by familial amyloid β protein precursor (βAPP$_{717}$) mutants. Science
264:1336–1340

Tamaoka A, Kalaria RN, Lieberburg I, Selkoe DJ (1992) Identification of a
stable fragment of the Alzheimer amyloid precursor containing the β-protein in brain microvessels. Proc Natl Acad Sci USA 89:1345–1349

Weidemann A, KÜnig G, Bunke D, Fischer P, Salbaum JM, Masters CL,
Beyreuther K (1989) Identification, biogenesis, and localization of precursors of Alzheimer's disease A4 amyloid protein. Cell 57:115–126

Yankner BA, Duffy LK, Kirschner DA (1990) Neurotrophic and neurotoxic
effects of amyloid β protein: reversal by tachykinin neuropeptides. Science
250:279–282

6 Regulation of Amyloid β-Protein Precursor Processing by Cell Surface Receptor Ligands and Second Messengers

R.M. Nitsch

6.1 Brain Amyloid as a Target for Drug Design

Deposition of brain amyloid plaques is an early and invariant neuropathological feature of Alzheimer's disease (AD). Up to 5%–20% of the brain cortex volume can be filled with amyloid in some forms of the disease (Hyman et al. 1995). Brain amyloid plaques are composed of insoluble aggregates of amyloid β peptide (Aβ) that is derived from proteolytic processing of the larger amyloid precursor protein (APP). APP is a member of a gene family that encodes ubiquitously expressed secretory glycoproteins (Kang et al. 1987; Slunt et al. 1994; Wasco et al. 1993; Weidemann et al. 1989) with yet unidentified functions related to cell adhesion, axogenesis, neurite outgrowth and neuroprotection (Barger et al. 1995; Milward et al. 1992; Ninomiya et al. 1995; Thinakaran et al. 1995; Williamson et al. 1995).

Several lines of evidence point to an involvement of APP in the pathogenesis of AD. Point mutations within the APP gene clustered around the region that encodes Aβ are linked to several familial forms of AD (Chartier-Harlin et al. 1991; Goate et al. 1991; Mullan et al. 1992; Murrell et al. 1991). Some of these disease-causing mutations are associated with changes in posttranslational processing of APP. In particular, the K670N-M671L mutation at the N-terminus of the Aβ domain linked to the "Swedish" variant of familial AD (Mullan et al. 1992) accelerates the generation of Aβ six- to eightfold as compared to the wild-type APP sequence (Cai et al. 1993; Citron et al. 1992, 1994). Furthermore, the V717F mutation at codon 717 close to the C-terminus of Aβ results in the generation of longer Aβ species (Suzuki et al. 1994). These are likely to aggregate more readily than wild-type species (Jarrett and Lansbury 1993), and they may therefore be associated with accelerated amyloid formation. Large quantities of brain amyloid deposits are invariantly present in brains of individuals with Down's syndrome at a young age, suggesting that the abnormally high expression of the APP gene due to the increased gene dosage of chromosome 21 genes can be associated with accelerated brain amyloid formation (Hyman et al. 1995; Rumble et al. 1989). This view is supported by recent transgenic experiments showing that overexpression of the human APP gene in the mouse can cause brain amyloid formation along with astrogliosis, reduced glucose utilization, seizures, and behavioral abnormalities (Games et al. 1995; Hsiao et al. 1995).

Familial AD cases linked to mutations of S182, a gene recently cloned on chromosome 14 (Alzheimer's Disease Collaborative Group 1995; Sherrington et al. 1995; Wasco et al. 1995) with many identified, distinct missense mutations, may further support a role of amyloid in the pathogenesis of AD. Cultured fibroblasts obtained from affected members of S182 AD families produce more APP mRNA (Querfurth et al. 1995) and secrete higher amounts of Aβ1–42(43) than those from nonaffected family members (Scheuner et al. 1995). Furthermore, S182 mutation carriers also have elevated plasma levels of Aβ1–42(43) (Song et al. 1995). Thus, a variety of genetic defects that cause the clinicopathological phenotype of AD are also associated with potentially increased amyloidogenicity of APP processing pathways. Together it seems reasonable to regard amyloid as a useful target for therapeutic interventions designed to reduce its formation, to inhibit its aggregation, and to accelerate its clearance from brain tissue.

6.2 Regulated APP Processing

Posttranslational processing of APP is characterized by maturation within the Golgi complex, packaging in secretory vesicles, translocation to the cell surface, and secretion of the ectodomain (APPs) after α-secretase cleavage within the Aβ domain (Esch et al. 1990; Sisodia et al. 1990; Weidemann et al. 1989). α-secretase cleavage occurs between positions 16 and 17 of the Aβ domain; it therefore presumably prevents the formation of Aβ. Alternatively, APP can be targeted to the endosomal compartment, either by reinternalization from the cell surface or directly from the Golgi complex. Within an endosomal compartment, it can be processed differentially to generate intact Aβ molecules that are also readily secreted (Haass et al. 1992; Shoji et al. 1992). As a result, both APPs and Aβ are present in conditioned tissue culture media, brain tissue (Gravina et al. 1995), as well as in extracellular fluids including human cerebrospinal fluid (CSF). CSF levels of APPs range between 1 and 3 μg/ml (1–3 nM), and levels of Aβ range between 10 and 30 ng/ml (2.3–7 nM) (Nitsch et al. 1996).

Secretory APP processing is highly regulated by a variety of internal and external signals, including protein kinase C (PKC), phospholipase A2, tyrosine kinases, and various neurotransmitters (Buxbaum et al. 1990; Emmerling et al. 1993; Nitsch et al. 1992; Nitsch and Growdon 1994; Slack et al. 1995). The APP processing pathways appear to be activity-dependent and may be modulated by neuronal activation (Farber et al. 1995; Nitsch et al. 1993), conceivably via a large family of G protein-coupled cell surface receptors that include acetylcholine, glutamate, serotonin, and peptide receptor subtypes (Lee et al. 1995; Nitsch et al. 1995; Nitsch et al. 1992).

6.3 Muscarinic, Serotonin, and Glutamate Receptors Regulate APP Processing

In cultured human 293 cells that stably overexpress muscarinic m1 or m3 receptor subtypes, stimulation with physiological or pharmacological ligands rapidly accelerates proteolytic cleavage of APP and increases the secretion of APPs into the cell culture media (Nitsch et al. 1992). Experiments with the translation-inhibiting agent cyclohexamide showed that receptor-mediated increase in APPs secretion occurred independently of protein synthesis and thus was not caused by possible receptor-mediated changes in APP gene expression or APP translation rates. Moreover, receptor stimulation reduced levels of cell-associated, full-length APP both in the presence and the absence of cycloheximide. Therefore, muscarinic receptor activation accelerates the proteolytic cleavage of preexisting, full-length APP, followed by the secretion of APPs. Receptor-mediated secreted APPs comprises the complete N-terminus of APP, including a substantial, but not the entire, portion of the Aβ N-terminus; this suggests that it is derived from APP cleavage within the Aβ domain, most likely by α-secretase cleavage between codons 16 and 17 of the Aβ domain.

To test directly whether muscarinic receptors can reduce Aβ formation, 293 cells were cotransfected with human wild-type APP or with APP comprising the K670N-M671L mutation. In these cells, both carbachol and phorbol myristate acetate (PMA) reduced the formation of soluble Aβ to approximately 40% of unstimulated control levels (Hung et al. 1993). Our results were confirmed in transfected African green monkey kidney COS and Chinese hamster ovary (CHO) cell lines, in human astrocytes and glioma cells, as well as in human NT2N neurons (Buxbaum et al. 1993; Gabuzda et al. 1993; Wolf et al. 1995). In cultured human neuroblastoma SY5Y cells, PMA-induced increases in APP secretion was associated with unchanged Aβ formation, suggesting that cell type-specific differences in the patterns of regulated APP processing are possible (Dyrks et al. 1994). By using pulse-chase experiments, we were able to show that muscarinic activation also reduced levels of several cell-associated C-terminal APP fragments of varying lengths ranging from 10 to 60 kDa. These observations imply that full-length APP is degraded more readily when cells are stimulated by the above external and internal factors.

Experiments with kinase inhibitors as well as overexpression and downregulation paradigms suggested a central role of PKC isoforms in the coupling of cell surface receptors and APP processing pathways (Slack et al. 1993), consistent with the idea that G-protein coupled m1 and m3 receptors can activate PKC via the PLC-mediated generation of diacylglycerol from phospholipid precursors (Nishizuka 1992).

Muscarinic m1 and m3 receptors are members of a larger superfamily of G protein-coupled cell-surface receptors that includes serotoninergic 5-HT2 and metabotropic glutamate receptor subtypes (Fisher et al. 1992). These receptors are widely distributed throughout the mammalian brain and are involved in neuronal functions such as long-term potentiation. We therefore tested their ability to regulate APP processing. By using the same subcloning and transfection strategy as described above, we found that the serotonin receptor subtypes 5-HT2a and 5-HT2c can also regulate APP processing (Nitsch et al. 1995). We observed that serotonin increased APPs release three- to fourfold from 3T3 cells stably overexpressing these receptors. The increase was dose dependent and was blocked by the serotonergic antagonists ketanserin, mianserin, and ritanserin. Activation of PKC with phorbol esters also increased APP secretion, but the protein kinase inhibitors chelerythrine chloride, and staurosporine failed to block the serotonin-induced increase in APPs secretion. Downregulation of PKC activity by prior long-term treatment with phorbol esters failed to inhibit the serotonin-induced increase in APP secretion, suggesting that activation of PKC is sufficient, but not necessary, to couple 5-HT receptor activation to increased APP secretion. 5-HT2aR-mediated increase in APP secretion was blocked by the phospholipase A_2 (PLA2) inhibitors manoalide, dimethyl-eicosadienoic acid (DEDA), and oleyloxyethyl phosphoryldioline (OPC), suggesting a role of PLA2 in the coupling of 5-HT2aR to secretory APP processing. In contrast, coupling of 5-HT2cR to APPs secretion involved both PKC and PLA2.

Serotonin also stimulated the release of the secretory form of the APP homolog APP-like protein 2 (APLP2), suggesting that members of the APP multigene family are processed via similar pathways. The serotoninergic agonist dexnorfenfluramine stimulated phosphatidylinositol turnover as well as the secretion of both APPs and APLP2s in 3T3 cells overexpressing 5-HT2aR or 5-HT2cR (Nitsch et al. 1995).

We were also able to show that metabotropic glutamate receptors are coupled to the regulation of APP processing. Metabotropic glutamate agonists stimulate APP secretion both in 293 cells transfected with the metabotropic glutamate receptor subtype mGluR1a and in primary rat hippocampal neurons (Lee et al. 1995). mGluR-mediated APP secretion was accompanied by increases in phosphatidylinositol turnover and was blocked by PKC inhibitors, suggesting that mGluR couple to APP processing via the PLC – PKC signaling pathway. In related experiments, we found that the peptides bradykinin and vasopressin also increased APPs secretion.

Together, these observations show that secretory APP processing can be regulated by multiple distinct external and internal signals. They suggest that the regulation of APP processing in brain is coupled to the activity of neuronal cells.

6.4 Regulated APP Processing in Mammalian Brain

In order to show directly that APP processing in mammalian brain is activity dependent, we used a tissue slice preparation freshly obtained from various regions of the rat brain, including hippocampus, cortex, striatum, and cerebellum (Farber et al. 1995; Nitsch et al. 1993). In this experimental system, neuronal activity can be accelerated by depolarization with electrical currents that induce action potentials along with the concomitant release of endogenous neurotransmitters from their presynaptic stores. These include acetylcholine, glutamate, and serotonin. The degree of neuronal activation in this system can be modulated by varying frequency, pulse duration, current density, and field strength. In addition to electrical depolarization paradigms, drugs can be tested by simply adding them to the superfusate media.

Electrical depolarization with physiological stimulation frequencies of 5–30 Hz at a pulse duration of 1 ms caused a frequency-dependent increase in the release of APPs (Nitsch et al. 1993). This increase was blocked by the sodium channel-inhibiting agent tetrodotoxin, indicating that neuronal action potentials can regulate APP processing in mammalian brain in an activity-dependent manner. In order to test whether cholinergic drugs can modulate APP processing pathways in the slice preparation, we tested the nonselective muscarinic agonist carbachol, the choline-esterase inhibitor physostigmine, the nonselective muscarinic antagonist atropine, the more selective m2 antagonist gallamine, and a selective, preferentially m1 agonist, WAL 2014 (Ensinger et al. 1993) for their ability to influence secretory APP processing in the brain slices. Atropine effectively blocked the stimulation-induced increase in APPs secretion, confirming that electrical stimulation of brain cells is coupled to regulated APP secretion via the stimulation of muscarinic receptors. Importantly, this blockade was only partial, suggesting the possibility that other neurotransmitter systems in addition to the muscarinic system may be involved in regulating APP processing in the slice preparation. Interestingly, both carbachol and physostigmine failed to modulate secretory APP processing. These drugs are nonselective compounds and cause the nonselective stimulation of all pre- and postsynaptic muscarinic receptor subtypes. In contrast, the more selective and preferentially m1 agonist WAL 2014 stimulates APP secretion to about twice the basal levels (Farber et al. 1995).

These data led to the hypothesis that nonselective activation of all muscarinic brain receptor subtypes blunts the m1-mediated acceleration of APP secretion. We tested this hypothesis by coadministering carbachol and gallamine, an m2 receptor antagonist. In confirmation of our hypothesis, this treatment caused more than twofold basal increases in APPs secretion (Farber et al. 1995). These data also imply that m2 receptor activation can interact with the m1 receptor-coupled increase in APP processing, either by blocking it or by reducing basal release of cells that are not affected by the m1 agonist. We therefore tested gallamine alone and found that it also accelerated rates of APP secretion, suggesting that basally released acetylcholine can effectively increase APP secretion when the presynaptic m2 receptor subtypes are blocked (Farber et al. 1995). It is also possible, however, that the gallamine-induced increase in basal acetylcholine was sufficient to accelerate APPs secretion. We are currently in the process of testing additional drugs in this brain slice system in order to determine which additional signaling systems regulate APP processing pathways in the brain.

6.5 Significance for Brain Physiology and Alzheimer's Disease

The physiological relevance of regulated APP secretion in the brain is unclear. We speculate that if APP processing is under neurotransmitter control, neuronal activity may regulate local tissue concentrations of a secretory molecule with possible paracrine activities related to neurotrophic and neuroprotective functions. Growth factor-like activities of APPs have been observed in several tissue culture models. In particular, a neurotrophin-like stimulation of neurite outgrowth and branching was observed in PC-12 cells (Milward et al. 1992). This activity may be related to the ability of APP to bind proteoglycans and other extracellular matrix proteins (Koo et al. 1993; Multhaup 1994; Small et al. 1994; Williamson et al. 1995). Conversely, antisense constructs directed against APP transcripts inhibit neurite outgrowth in primary neurons (Allinquant et al. 1995), underscoring a role of APP in neurite growth.

Regulated cleavage of a membrane precursor with growth-promoting properties followed by secretion of the ectodomain was described for a variety of transmembrane proteins, including transforming growth factor-α (TGF-α) (Bosenberg et al. 1992; Pandiella and Massague 1991) and the tumor necrosis factor (TNF) receptor (Brakebusch et al. 1992). It was proposed that regulated secretion is involved in switching TGF-α activity from that of a juxtacrine to a paracrine growth factor (Pandiella and Massagure 1991). In analogy, secreted APPs and APLPs may be regulated growth factors with trophic functions unrelated to APP's role as an amyloid precursor. Activity-dependent release of these trophic molecules may enhance the contact of actively used synapses and promote neurite outgrowth of firing neurons.

Reduced neurotransmission and neuronal degeneration in AD brain may contribute to misprocessing of APP. Reduced α-secretase processing may accelerate the formation of amyloidogenic derivatives and promote amyloid formation. Amyloid deposits in AD brain are clearly present throughout the entire brain cortex, and they are not colocalized with any specific neurotransmitter system. Similarly, many neurotransmitter systems including the cholinergic, serotoninergic, glutamatergic, and peptidergic systems are heavily damaged in AD brains (Bowen et al. 1989; Francis et al. 1993). This damage is associated with significant dysfunction and losses in cortical synapses (Terry et al. 1991). These pathological alterations along with the resulting deafferentation of target cells may be associated with abnormal APP processing and increased formation of amyloidogenic APP derivatives.

It is important to note, however, that neurotransmission is defective in many other brain diseases that are not associated with increased amyloid formation. These include Parkinson's disease, Huntington's disease, and many system atrophies such as olivo-ponto-cerebellar atrophy, and corticostriatal degeneration. Moreover, amyloid is deposited throughout the neuropil, also in brain areas that are otherwise relatively spared from neurodegeneration (e.g., cerebellum). Thus, additional factors may be contributing to brain amyloid formation.

6.6 Receptor Ligands as Pharmacological Tools to Regulate APP Processing

Identifying cell surface receptors, such as m1, m3, 5-HT2aR, 5-HT2cR, and mGluR1α, whose stimulation increases APPs secretion, could constitute a useful novel pharmacological strategy for manipulating APP processing in brain. Ligands for muscarinic, serotonin, and glutamate receptors may promote the physiological functions of APP as a regulated paracrine neuronal growth factor. Concomitantly, compounds that increase α-secretase processing may reduce tissue levels of amyloidogenic APP derivatives. Inasmuch as aggregation of soluble Aβ is a function of its concentration, decreased concentrations may preclude an essential and initial step in amyloid formation. Secondly, reducing the rate at which Aβ is produced may slow the growth of preexisting brain amyloid plaques.

With the successful generation of transgenic mice that develop AD-type brain amyloid deposits, this hypothesis can now be tested under in vivo conditions by measuring amyloid formation in these animals in response to treatments with receptor agonists designed to reduce Aβ formation.

If confirmed in vivo, clinical trials will be needed to determine whether subtype-selective receptor agonists can also ameliorate the clinical symptoms of AD. Moreover, clinical studies are necessary to determine whether preventive strategies that commence the treatment with Aβ-reducing drugs before the clinical onset of AD can reduce the risk of getting AD, delay its onset, and slow the clinical course of the disease.

References

Allinquant B, Hantraye P, Mailleux P, Moya K, Bouillot C, Prochiantz (1995) Downregulation of amyloid precursor protein inhibits neurite outgrowth in vitro. J Cell Biol 128:919–927

Alzheimer's Disease Collaborative Group (1995) The structure of the presenilin 1 (S182) gene and identification of six novel mutations in early onset Alzheimer's disease. Nature Genet 2:219–222

Barger SW, Fiscus RR, Ruth P, Hofmann F, Mattson MP (1995) The role of cyclic GMP in the regulation of neuronal calcium and survival by secreted forms of β-amyloid precursor. J Neurochem 64:2087–2096

Bosenberg MW, Pandiella A, Massague J (1992) The cytoplasmic carboxy-terminal amino acid specifies cleavage of membrane TGF α into soluble growth factor. Cell 71:1157–1165

Bowen DM, Najlerahim A, Procter AW, Francis PT, Murphy E (1989) Circumscribed changes of the cerebral cortex in neuropsychiatric disorders of later life. Proc Natl Acad Sci USA 86:9504–9508

Brakebusch C, Nophar Y, Kemper O, Engelmann H, Wallach D (1992) Cytoplasmic truncation of the p55 tumor necrosis factor (TNF) receptor abolishes signalling, but not induced shedding of the receptor. EMBO J 11:943–950

Buxbaum JD, Gandy SE, Cicchetti P, Ehrlich ME, Czernik AJ, Fracasso RP, Ramabhadran TV, Unterbeck AJ, Greengard P (1990) Processing of Alzheimer βA4 amyloid precursor protein: modulation by agents that regulate phosphorylation. Proc Natl Acad Sci USA 87:6003–6006

Buxbaum JD, Koo EH, Greengard P (1993) Protein phosphorylation inhibits production of Alzheimer amyloid β/A4 peptide. Proc Natl Acad Sci USA 90: 9195–9198

Cai X-D, Golde TE, Younkin SG (1993) Release of excess amyloid β protein from a mutant amyloid β protein precursor. Science 259:514–516

Chartier-Harlin M-C, Crawford F, Houlden H (1991) Early-onset Alzheimer's disease caused by mutations at codon 717 of the β-amyloid precursor protein gene. Nature 353:844–846

Citron M, Oltersdorf T, Haass C, McConlogue L, Hung AY, Seubert P, Vigo-Pelfrey C, Lieberburg I, Selkoe DJ (1992) Mutation of the β-amyloid precursor protein in familial Alzheimer's disease increases β-protein production. Nature 360:672–674

Citron M, Vigo-Pelfrey C, Teplow DB, Miller C, Schenk D, Johnston J, Winblad B, Venizelos N, Lannfelt L, Selkoe DJ (1994) Excessive production of amyloid β-protein by peripheral cells of symptomatic and presymptomatic patients carrying the Swedish familial Alzheimer's disease mutation. Proc Natl Acad Sci USA 91:11993–11997

Dyrks T, Msnning U, Beyreuther K, Turner J (1994) Amyloid precursor protein secretion and beta A4 amyloid generation are not mutually exclusive. FEBS Lett 349:210–214

Emmerling MR, Moore CJ, Doyle PD, Carroll RT, Davis RE (1993) Phospholipase A2 activation influences the processing and secretion of the amyloid precursor protein. Biochem Biophys Res Commun 197:292–297

Ensinger HA, Doods HN, Immel-Sher AR, Kuhn FJ, Lambrecht G, Mendla KD, Muller RE, Mutschler E, Sagrada A, Walther G, Hammer R (1993) WAL 2014 – a muscarinic agonist with preferential neuron-stimulating properties. Life Sci 52:473–480

Esch FS, Keim PS, Beattie EC, Blacher RW, Culwell AR, Oltersdorf T, McClure D, Ward PJ (1990) Cleavage of amyloid β peptide during constitutive processing of its precursor. Science 248: 1122–1124

Farber SA, Nitsch RM, Schulz JG, Wurtman RJ (1995) Regulated secretion of β-amyloid precursor protein in rat brain. J Neurosci 15:7442–7451

Fisher SK, Heakock AM, Agranoff BW (1992) Inositol phospholipids and signal transduction in the nervous system: an update. J Neurochem 58:18–38

Francis PT, Sims NR, Procter AW, Bowen DM (1993) Cortical pyramidal neurone loss may cause glutamatergic hypoactivity and cognitive impairment in Alzheimer's disease: investigative and therapeutic perspectives. J Neurochem 60:1589–1604

Gabuzda D, Busciglio J, Yankner BA (1993) Inhibition of β-amyloid production by activation of protein kinase C. J Neurochem 61:2326–2329

Games D, Adams D, Alessandrini R, Barbour R, Berthelette P, Blackwell C, Carr T, Clemens J, Donaldson T, Gillespie F, Guido T, Hagopian S, Johnson-Wood K, Khan K, Lee M, Leibowitz P, Lieberburg I, Little S, Masliah E, McConlogue L, Montoya-Zavala M, Mucke L, Paganini L, Penniman E, Power M, Schenk D, Seubert P, Snyder B, Soriano F, Tan H, Vitale J, Wadsworth S, Wolozin B, Zhao J (1995) Alzheimer-type neuropathology in transgenic mice overexpressing V717F β-aymloid precursor protein. Nature 373:523–527

Goate A, Chartier-Harlin M-C, Mullan M, Broen J, Crawford F, Fidani L, Giuffra L, Hayes A, Irving N, James L, Mant R, Newton P, Rooke K, Roques P, Talbot C, Pericak-Vance M, Roses A, Williamson R, Rossor M, Owen M, Hardy J (1991) Segregation of a missense mutation in the amyloid precursor gene with familial Alzheimer's disease. Nature 349:704–706

Gravina SA, Ho L, Eckman CB, Long KE, Otvos LJ, Younkin LH, Suzuki N, Younkin SG (1995) Amyloid β protein (Aβ) in Alzheimer's disease brain. J Biol Chem 270:7013–7016

Haass C, Schlossmacher MG, Hung AY, Vigo-Pelfrey C, Mellon A, Ostaszewski BL, Lieberburg I, Koo EH, Schenk D, Teplow DB, Selkoe DJ

(1992) Amyloid β-peptide is produced by cultured cells during normal metabolism. Nature 359: 322–325

Hsiao KK, Borchelt DR, Olson K, Johannsdottir R, Kitt C, Yunis W, Xu S, Eckman C, Younkin S, Price D, Iadecola C, Clark HB (1995) Age-related CNS disorder and early death in transgenic FVB/N mice overexpressing Alzheimer amyloid precursor proteins. Neuron 15:1203–1218

Hung AY, Haass C, Nitsch RM, Qiao Qiu W, Citron M, Wurtman RJ, Growdon JH, Selkoe DJ (1993) Activation of protein kinase C inhibits cellular production of the amyloid β-protein. J Biol Chem 268:22959–22962

Hyman BT, West HL, Rebeck GW, Buldyrev SV, Mantegna RN, Ukleja M, Havlin S, Stanley HE (1995) Quantitative analysis of senile plaques in Alzheimer's disease: observation of log-normal size distribution and molecular epidemiology of differences associated with apolipoprotein E genotape and trisomy 21 (Down syndrome). Proc Natl Acad Sci USA 92:3586–3590

Jarrett JT, Lansbury PT Jr (1993) Seeding "one-dimensional crystallization" of amyloid: a pathogenic mechanism in Alzheimer's disease and Scrapie? Cell 73:1055–1058

Kang J, Lemaire H-G, Unterbeck A, Salbaum JM, Masters CL, Grzeschik KH, Multhaup G, Beyreuther K, Müller-Hill B (1987) The precursor of Alzheimer's disease amyloid A4 protein resembles a cell-surface receptor. Nature 325:733–736

Koo EH, Park L, Selkoe DJ (1993) Amyloid β-protein as a substrate interacts with extracellular matrix to promote neurite outgrowth. Proc Natl Acad Sci USA 90:4748–4752

Lee RKK, Wurtman RJ, Slack BE, Cox AJ, Nitsch RM (1995) Amyloid precursor protein processing is stimulated by metabotropic glutamate receptors. Proc Natl Acad Sci USA 92:8083–8087

Milward EA, Papadopoulos R, Fuller SJ, Moir RD, Small D, Beyreuther K, Masters CL (1992) The amyloid protein precursor of Alzheimer's disease is a mediator of the effects of nerve growth factor on neurite outgrowth. Neuron 9:129–137

Mullan M, Crawford F, Axelman K, Houlden H, Lilius L, Winblad B, Lannfelt L (1992) A pathogenic mutation for probable Alzheimer's disease in the APP gene at the N-terminus of beta-amyloid. Nature Genet 1:345–347

Multhaup G (1994) Identification and regulation of the high affinity binding site of the Alzheimer's disease amyloid protein precursor (APP) to glycosaminoglycans. Biochemie 76:304–311

Murrell J, Farlow M, Ghetti B, Benson MD (1991) A mutation in the amyloid precursor protein associated with hereditary Alzheimer's disease. Science 254:97–99

Ninomiya H, Roch JM, Jin LW, Saitoh T (1995) Secreted forms of amyloid β/A4 protein precursor (APP) binds to two distinct APP binding sites on rat

B102 neuron-like cells through two different domains, but only one site is involved in neuritotrophic activity. J Neurochem 63:495–500

Nishizuka Y (1992) Intracellular signaling by hydrolysis of phospholipids and activation of protein kinase C. Science 258:607–614

Nitsch RM, Growdon JH (1994) Role of neurotransmission in the regulation of amyloid β-protein precursor processing. Biochem Pharmacol 47:1275–1284

Nitsch RM, Slack BE, Wurtman RJ, Growdon JH (1992) Release of Alzheimer amyloid precursor derivatives stimulated by activation of muscarinic acetylcholine receptors. Science 258:304–307

Nitsch RM, Farber SA, Growdon JH, Wurtman RJ (1993) Release of amyloid β-protein precursor derivatives from hippocampal slices by electrical depolarization. Proc Natl Acad Sci USA 90:5191–5193

Nitsch RM, Rebeck GW, Deng M, Richardson UI, Tennis M, Schenk DB, Vigo-Pelfrey C, Lieberburg I, Wurtman RJ, Hyman BT, Growdon JH (1995) Cerebrospinal fluid levels of amyloid β-protein in Alzheimer's disease: inverse correlation with severity of dementia and effect of apolipoprotein genotype. Ann Neurol 37:512–518

Nitsch RM, Deng M, Growdon JH, Wurtman RJ (1996) Serotonin 5-HT2a and 5-HT2c receptors stimulate APPs secretion. J Biol Chem 271:4188–4194

Pandiella A, Massague J (1991) Cleavage of the membrane precursor for transforming growth factor alpha is a regulated process. Proc Natl Acad Sci USA 88:1726–1730

Querfurth HW, Wijsman EM, St. George-Hyslop PH, Selkoe DJ (1995) bAPP mRNA transcrition in increased in cultured fibroblasts from the familial Alzheimer's disease-1 family. Mol Brain Res 28:319–337

Rumble B, Retallack R, Hilbich C, Simms G, Multhaup G, Martins R, Hockey A, Montgomery P, Beyreuther K, Masters CL (1989) Amyloid A4 protein and its precursor in Down's syndrome and Alzheimer's disease. N Engl J Med 320:1446–1452

Scheuner D, Bird T, Citron M, Lannfelt L, Schellenberg G, Selkoe D, Viitanen M, Younkin SG (1995) Fibroblasts from carriers of familial AD linked to chromosome 14 show increased Aβ production. Soc Neurosci Abstr 21:1500

Sherrington R, Rogaev EI, Liang Y, Rogaeva EA, Levesque G, Ikeda M, Chi H, Lin C, Li G, Holman K, Tsuda T, Mar L, Foncin J-F, Bruni AC, Montesi MP, Sorbi S, Rainero I, Pinessi L, Nee L, Chumakov I, Pollen D, Brookes A, Sanseau P, Polinsky RJ, Wasco W, Da Silva HAR, Haines JL, Pericak-Vance MA, Tanzi RE, Roses AD, Fraser PE, Rommens JM, St George-Hyslop PH (1995) Cloning of a gene bearing missense mutations in early-onset familial Alzheimer's disease. Nature 375:754–760

Shoji M, Golde TE, Ghiso J, Cheung TT, Estus S, Shaffer LM, Cai X-D, McKay DM, Tintner R, Frangione B, Younkin SG (1992) Production of the Alzheimer amyloid β-protein by normal proteolytic processing. Science 258:126–129

Sisodia SS, Koo EH, Beyreuther K, Unterbeck A, Price DL (1990) Evidence that β-amyloid protein of Alzheimer's disease is not derived by normal processing. Science 248:492–495

Slack BE, Nitsch RM, Livneh E, Kunz GM Jr, Breu J, Eldar H, Wurtman RJ (1993) Regulation by phorbol esters of amyloid precursor protein release from Swiss 3T3 fibroblasts overexpressing protein kinase Cα. J Biol Chem 268:21097–21101

Slack BE, Breu J, Petryniak MA, Srivastava K, Wurtman RJ (1995) Tyrosine phosphorylation-dependent stimulation of amyloid precursor protein secretion by the m3 muscarinic acetylcholine receptor. J Biol Chem 270:8337–8344

Slunt HH, Thinakaran G, von Koch C, Lo ACY, Tanzi RE, Sisodia SS (1994) Expression of a ubiquitous, cross-reactive homologue of the mouse β-amyloid precursor protein (APP). J Biol Chem 269:2637–2644

Small DH, Nurcombe V, Reed G, Clarris H, Moir R, Beyreuther K, Masters C (1994) A hepatin-binding domain in the amyloid protein precursor of Alzheimer's disease is involved in the regulation of neurite outgrowth. J Neurosci 14:2117–2127

Song X-H, Suzuki N, Bird T, Peskind E, Schellenberg GY (1995) Plasma amyloid β protein (Aβ) ending at Aβ42(43) is increased in carriers of familial AD (FAD) linked to chromosome 14. Soc Neurosci Abstr 21:1501

Suzuki N, Cheung TT, Cai X-D, Odaka A, Otvos L Jr, Echman C, Golde TE, Younkin SG (1994) An increased percantage of long amyloid β protein secreted by familial amyloid β protein precursor (βAPP717) mutants. Science 264:1336–1340

Terry RD, Masliah E, Salmon DP, Butters N, DeTeresa R, Hill R, Hansen LA, Katzman R (1991) Physical basis of cognitive alterations in Alzheimer's disease: synapse loss is the major correlate of cognitive impairment. Ann Neurol 30:572–580

Thinakaran G, Kitt CA, Roskams AJI, Slunt HH, Masliah E, von Koch C, Ginsberg SD, Ronnett GV, Reed RR, Price DL, Sisodia SS (1995) Distribution of an APP homolog, APLPs in the mouse olfactory system: a potential role for APLP2 in axogenesis. J Neurosci 15:6314–6326

Wasco W, Gurubhagavatula SS, Paradis MD, Romano DM, Sisodia SS, Hyman BT, Neve RL, Tanzi RE (1993) Isolation and characterization of APLP2 encoding a homologue of the Alzheimer's associated amyloid β protein precursor. Nature Genet 5:95–99

Wasco W, Pettingell WP, Jondro PD, Schmidt SD, Gurubhagavatula S, Rodes L, DiBlast T, Romano DM, Guenette SY, Kovacs DM, Growdon JH, Tanzi RE (1995) Familial Alzheimer's disease chromosome 14 mutations. Nature Med 1:848

Weidemann A, Ksnig G, Bunke D, Fischer P, Salbaum JM, Masters CL, Beyreuther K (1989) Identification, biogenesis, and localization of precursors of Alzheimer's disease A4 amyloid protein. Cell 57:115–126

Williamson TG, Nurcombe V, Beyreuther K, Masters CL, Small DH (1995) Affinity purification of proteoglycans that bind to the amyloid protein precursor of Alzheimer's disease. J Neurochem 65:2201–2208

Wolf BA, Wertkin AM, Jolly YC, Yasuda RP, Wolfe BB, Konrad RJ, Manning D, Ravi S, Williamson JR, Lee VMY (1995) Muscarinic regulation of Alzheimer's disease amyloid precursor protein secretion and amyloid beta-protein prduction in human neuronal NT2N cells. J Biol Chem 270:4916–4922

7 Tau Phosphorylation and Dephosphorylation in the Pathogenesis of Alzheimer's Disease

M. Mawal-Dewan, J.Q. Trojanowski, and V.M.-Y. Lee

7.1 Introduction

Alzheimer's disease (AD) is the most common dementing illness among individuals over 60 years old, the most rapidly growing segment of the population in developed countries. The AD brain contains two conspicuous types of deposits, the extracellular senile or amyloid plaques and the intraneuronal neurofibrillary tangles (NFTs). NFTs and senile plaques (SPs) were first described by Alois Alzheimer in 1907 and their presence in large numbers has since been recognized as one of the major neuropathological characteristics of the disease that bears his name. NFTs are formed within nerve cells that degenerate during the course of the disease where they are found in the cell body, and related neurofibrillary lesions also are seen in abnormal neurites associated with amyloid plaques and within the neuropil threads (Alzheimer 1907;

Braak et al. 1986). The relative insolubility of NFTs enables their survival after the death of affected nerve cells as extracellular NFTs (or ghost cells) that accumulate in the neuropil (Alzheimer 1907; Braak et al. 1986; Kidd 1963). These extracellular NFTs are probably then engulfed by astrocytes and/or microglial cells where they are presumably slowly degraded. Amyloid plaques consist of aggregates of amyloid peptides (Aβ) which are derived from transmembrane proteins (APP, amyloid precursor protein) as a result of normal proteolysis (Kosik 1992). NFTs and their constituents, the paired helical filaments (PHFs), consist largely of the microtubule-associated protein tau in an abnormal state of phosphorylation (Lee et al. 1993a). These deposits are particularly suitable for classifying the stages of AD progress since the number and location of the neurofibrillary changes are closely related to the severity of the disease (Braak and Braak 1991). Although NFTs and related neurofibrillary lesions are not restricted to AD (Joachim and Selkoe 1992; Tagliavani et al. 1993; Trojanowski et al. 1990, 1993a), recent studies have confirmed a close correlation between the burden of neurofibrillary abnormalities and the presence of dementia or the severity of the cognitive impairments (Arriagada et al. 1992; Dickson et al. 1991). Consequently, neurofibrillary lesions have been the focus of intense efforts to gain insights into how neurons become dysfunctional and degenerate in AD.

7.2 Tau Structure

Normal tau protein is abundant in both the central (CNS) and peripheral nervous systems (PNS) (Goedert et al. 1994a), where it stabilizes microtubules (Cleveland et al. 1977). It consists of at least six tau isoforms in human brain tissue generated by alternate splicing of a single gene (Goedert et al. 1989a,b). These isoforms range from 352 to 441 amino acids in length and they differ from each other by the presence or absence of different amino or carboxy terminal inserts (Goedert et al. 1989a; Goedert and Jakes 1990). The most striking feature of the structure of tau is stretches of 31 or 32 amino acids, repeated three or four times in the carboxy terminal half or in the microtubule (MT) binding units. The function of these amino terminal inserts is unknown. The shortest isoform is comprised of 352 amino acids and it contains three repeats and no amino terminal inserts, whereas the largest isoform is 441 amino acids in length and contains four repeats and the 58 amino acid insertion located near the amino terminus. The expression of these alternatively spliced tau isoforms is developmentally regulated in the human (Goedert 1993; Goedert et al. 1989a) and rat (Mawal-Dewan et al. 1994) CNS. While all six of these isoforms are found in PHFs, another isoform with a large additional insert in the amino terminal half described in the PNS (Goedert et al. 1992a; Couchie et al. 1992) as "big-tau" has not been found in PHFs.

7.3 Hyperphosphorylated Tau Proteins Are the Building Blocks of PHF-Tau

Although all six normal human tau isoforms are phosphoproteins, the tau in PHFs (PHF-tau) has been reported to be "hyperphosphorylated" or "abnormally" phosphorylated on sites that are not found in normal adult human tau isolated from postmortem brain samples (Lee et al. 1991; Hasagawa, et al. 1992; Goedert 1993). These and other studies (Goedert et al. 1993; Trojanowski et al. 1993a,b) have led to the hypothesis that the excessive or abnormal phosphorylation of tau is the key event in the transformation of normal adult tau into PHF-tau during the progression of AD.

Earlier controversy existed over the role of tau in PHF formation because another candidate PHF protein known as the ALZ50 antigen (or A68) was shown to have biochemical properties distinct from normal adult human tau (Greenberg and Davies 1990; Ksiezak-Reding et al. 1990; Vincent and Davies 1990; Liu et al. 1991). This confusion was resolved when A68 was purified from AD brains and shown to contain PHFs and to consist of an abnormally phosphorylated (e.g., at Ser-396) variant of tau (Bramblett et al. 1992, 1993; Greenberg et al. 1992; Khatoon et al. 1992; Lee et al. 1991; Goedert et al. 1992b). These findings have been extended (Brion et al. 1991; Greenberg et al. 1992; Ksiezak-Reding and Yen 1991) and recombinant tau fragments, including the four MT-binding repeats also have been induced to form PHFs in vitro (Crowther et al. 1992; Wille et al. 1992). Moreover, the isolation of tau fragments from purified PHFs added credence to the notion that tau proteins are the subunits of PHF (Goedert et al. 1988; Wischik et al.

1988). Additionally, in situ epitope analysis of tau protein domains in NFTs and in isolated PHFs revealed antigenic determinants spanning nearly the entire length of tau (Kosik et al. 1988). Taken together, these data provide compelling evidence that PHFs isolated from AD brains are comprised of A68 or PHF-tau, i.e., abnormally phosphorylated tau isoforms.

Recent studies have identified a number of specific phosphorylation sites in normal human fetal and adult tau as well as in PHF-tau isolated from postmortem brain samples, in addition to fetal and adult rat tau isolated from fresh brain samples (Hasegawa et al. 1992; Watanabe et al. 1993). These studies used direct chemical (Poulter et al. 1993; Watanabe et al. 1993) analysis and phosphorylation-dependent anti-tau antibodies (Kanemura et al. 1992; Goedert et al. 1993, 1994a,b; Goedert et al. 1995; Bramblett et al. 1993; Kenessey and Yen 1993; Brion et al. 1993; Hasegawa et al. 1993; Matsuo et al. 1994; Seubert et al. 1995). Thus in fetal tau, tau is phosphorylated at serine residues 198, 199, 202, 235, 262, 396, 400, and 404 and at threonine residues 181, 205, 217 and 231 (according to the numbering of the longest human brain tau iso-form). In adult brain, tau is phosphorylated at serine residues 199, 202, 235, 262, 396, and 404 and at threonine residues 181, 205, and 231. However, PHF-tau has been shown to be phosphorylated at serine residues 198, 199, 202, 208, 210, 214, 231, 235, 262, 396, 400, 404, 409, 412, 413, and 422 and threonine residues 181, 205, 212, 217, and 403. Some of these sites are also phosphorylated in tau from developing rat brain (Mawal-Dewan et al. 1994). Although similar sites of phos-phorylation were not detected in tau from autopsied adult human brain, recent studies ofadult human brain tau obtained from fresh cortical biopsy samples suggest that the majority of these sites are also phospho-rylated in adult human brain (Matsuo et al. 1994) albeit to a smaller extent than in fetal brain or PHF-tau. However a number of additional sites have been identified as phosphorylation sites in PHF-tau by mass spectroscopy that have so far not been found in rapidly isolated fetal rat tau or in biopsy-derived adult human brain tau (Morishima-Kawashima et al. 1995). It remains to be seen whether these sites are truely abnor-mally phosphorylated in PHF-tau. This will depend on the production of novel phosphorylation-dependent anti-tau antibodies and on future mass spectroscopic studies performed on rapidly isolated fetal and adult human brain tau. With the exception of Ser-262 all of the known phosphorylation sites in tau are located outside the microtubule-binding repeat region. Phosphorylation is heterogeneous, implying that a given tau molecule is phosphorylated at some, but not all, of these sites. Many of these are Ser/Thr residues that are followed by a proline, suggesting that protein kinases with a specificity for seryl-proline and threonyl-proline phosphorylate tau in normal brain. Accordingly, mitogen-activated protein kinases (MAP; Goedert et al 1992c; Drewes et al. 1992; Ledesma et al. 1992; Trojanowski et al. 1993c), glycogen synthase kinase-3 (GSK-3; Hanger et al. 1992; Mandelkow et al. 1992); cyclin-dependent kinase 5 (cdk-5; Paudel et al. 1993; Kobayashi et al. 1993; Baumann et al. 1993) and cdc-like kinase (Mawal-Dewan et al. 1992; Vulliet et al. 1992) phosphorylate tau at a number of the above Ser/Thr-Pro residues in vitro. In addition, cAMP-dependent protein kinase and Ca^{2+}/calmodulin-dependent protein kinase II phosphorylate serine residue 262 of tau in vitro (Litersky et al. 1995), as does a partially purified 110 kDa protein kinase from brain (Drewes et al. 1995). A very recent reassessment of the sites of phosphorylation in PHF-tau using ion spray mass spectroscopy led to the identification of 19 such sites in PHF-tau and 10 of these sites are found to be non-proline-directed (Morishima-Kawashima et al. 1995). Since there are as many as 17 Ser/Pro or Thr/Pro sites in the largest human brain tau isoform and 14 of these sites are shared by all 6 CNS tau proteins, it will be important to verify the identity of those proline-directed and nondirected sites that indeed are aberrantly phosphorylated in PHF-tau relative to normal fetal and adult human CNS tau. Thus, the role of these kinases in regulating the phosphorylation state of tau in vivo remains to be clarified. Finally, recent studies suggest that the protein tyrosine kinase *fyn* may regulate the activity of candidate proline-directed kinases involved in the abnormal phosphorylation of CNS tau, and that *fyn* thus may be involved in abnormal phosphorylation cascades in the AD brain (Shirazi and Wood 1993). Indeed, recent studies demonstrating that GSK 3α and GSK 3β but not MAP kinase can phosphorylate tau at specific sites in transfected COS-1 or CHO cells expressing tau isoforms, respectively (Lodestone et al. 1994; Sperber et al., unpublished data), emphasize the importance of in vivo studies for the ultimate identification of kinases that phosphorylate tau in neurons.

   The phosphorylation state of a protein is a result of a balance between protein kinase and protein phosphatase activities. Previous studies have demonstrated that the Ser/Thr protein phosphatases PP2A1 and PP2B (calcineurin) can dephosphorylate tau in vitro (Harris et al. 1990 Goedert et al. 1992c; Drewes et al. 1993). Furthermore, our recent study of biopsy-derived human tau also implicates PP2A as a phosphatase involved in dephosphorylating tau. The most plausible explanation for the failure to detect phosphorylation in adult CNS tau at several sites previously identified in PHF-tau and fetal tau is the rapid dephosphorylation of adult tau in brain tissue samples during brief postsurgical or postmortem delays in tissue processing to isolate tau (Matsuo et al. 1994; Garver et al. 1994). This increase in the dephosphorylation of tau could be due to either the constitutive activity of phosphatases or a net increase in the activity of phosphatases relative to kinases in brain tissue during postmortem or postsurgical delays. This increase in dephosphorylation appeared to be due to the constitutive activation of phosphatase as suggested by studies of rat brain tau isolated from brains of different developmental ages (Mawal-Dewan et al. 1994). Tau was isolated using the classical method of MT reassembly in the presence or absence of the phosphatase inhibitor okadaic acid (Mawal-Dewan et al. 1994) and these studies showed that fetal or neonatal brain tau isolated with or without phosphatase inhibitors did not differ substantially in their extent of phosphorylation. In contrast, adult rat brain tau isolated in the absence of phosphatase inhibitors was much more dephosphorylated at many sites when compared to adult brain tau isolated in the presence of phosphatase inhibitors. These studies indicate that there is an increase in phosphatase activities which are capable of dephosphorylating tau in the adult brain. However, it is at present unclear whether the underlying defect in the AD brain results from an increased protein kinase activity or a decreased protein phosphatase activity or from a combination of both. The finding that nonassembled soluble tau is much more prone to postmortem dephosphorylation than assembled PHF-tau has led to the suggestion that a reduced activity of protein phosphatases may be responsible for the hyperphosphorylation of tau in the AD brain (Matsuo et al. 1994; Mawal-Dewan et al. 1994). However, additional efforts are needed to dissect out the complex interplay of a potentially large array of protein kinases and protein phosphatases that may be involved in the progressive conversion of normal tau into PHF-tau.

7.4 Effect of Tau Phosphorylation
on Tau–Microtubule Interaction

The affinity of the binding of tau to MTs is modulated by phosphorylation as well as by the developmentally regulated increase in the number of MT-binding repeats. Fetal forms of rat and human tau have only three repeats with no amino terminal inserts, and they do not bind to MTs as well as tau isoforms containing four repeats (Goedert and Jakes 1990; Butner and Kirschner 1991). During development, human fetal tau is more phosphorylated than adult tau. Thus, the reduced number of MT-binding repeats and the increased phosphorylation of fetal tau would result in a decreased affinity of fetal tau for MTs, allowing the cytoskeleton to reorganize as immature axons grow and establish synapses with their proper targets. Comparisons of PHF-tau with adult and fetal tau proteins have demonstrated that PHF-tau recapitulates the phosphorylation state of fetal CNS tau (Bramblett et al. 1993; Goedert et al. 1993; Kanemura et al. 1992; Lee et al. 1993). For example, Ser-202 and Ser-396 are phosphorylated in PHF-tau and in the smallest tau isoform when it is expressed in the fetal CNS, but not in any of the six postmortem-derived tau isoforms in the normal adult CNS including the smallest tau protein that is present in both the fetal and adult brain (Bramblett et al. 1993; Goedert et al. 1993). Despite the fact that fetal CNS tau is phosphorylated at Ser-202 and Ser-396 as in PHF-tau, fetal tau is capable of binding to MTs, whereas PHF-tau completely loses the ability to bind MTs (Bramblett et al. 1993). The reduced binding of PHF-tau to MTs, coupled with reduced levels of normal tau in the AD brain, probably destabilizes MTs during the progression of AD, resulting in the impairment of vital cellular processes, such as rapid axonal transport leading to the degeneration of affected nerve cells. However, the normal programmed death of large numbers of neurons in the developing CNS occurs without any of the neurofibrillary lesions like those found in the brain in AD (Lee et al. 1993; Tohyama et al. 1991; Yachnis et al. 1993). Taken together, these observations suggest that the accumulation of neurofibrillary lesions in AD may be due to aberrant reactivation of fetal protein kinases and the inactivation of protein phosphatases in the AD brain that normally determine the phosphorylation state of fetal tau in the developing brain. To understand this issue better, it is important to understand how the phosphorylation state of tau is regulated in the fetal and adult human brain.

Previous studies have shown that the reversible phosphorylation and dephosphorylation of tau regulates its binding to MTs (Bramblett et al. 1993; Biernat et al. 1993; Linwall and Cole 1984; Dreschel et al. 1992). PHF-tau is much more extensively phosphorylated than rapidly processed normal adult human CNS tau proteins, and PHF-tau does not bind to MTs unless it is dephosphorylated (Bramblett et al. 1993; Yoshida and Ihara 1993). Hence, studies conducted in recent years have focused on determining the contribution of different phosphorylation sites in regulating the binding of tau to MTs. Indeed, some of the phosphorylation sites have been suggested to play a critical role in regulating the binding of tau to MTs. For example, in vitro MT binding studies using phosphorylated and nonphosphorylated recombinant wild-type and mutant tau proteins (Biernat et al. 1993; Drewes et al. 1995) have emphasized the critical role of the phosphorylation of Ser-262 (which resides in the first MT-binding repeat of tau) as the dominant regulator of the binding of tau to MTs. In contrast, other studies have implicated those phosphate acceptor sites preceded by a proline that flank the MT-binding domains (e.g., Ser-396) in regulating the binding of tau to MTs (Bramblett et al. 1993; Dreschel et al 1992). However, Seubert et al. (1995) have recently demonstrated that Ser-262 is a normal site of phosphorylation in human fetal and adult CNS tau in addition to PHF-tau, and that adult brain tau phosphorylated at Ser-262 can bind to MTs since it is recovered from the high-spin supernatant with MTs reassembled from brain extracts. Thus, these findings suggest that phosphorylation of Ser-262 alone is not sufficient to eliminate the binding of tau to MTs (Seubert et al 1995). Since the phosphorylation of tau at proline-directed sites flanking the MT-binding repeats (i.e., Ser-396/Ser-404 and Ser-202/Thr-205) can also be recovered bound to MTs, these sites do not independently regulate the binding of tau to MTs. Instead, the simultaneous phosphorylation of tau at multiple sites may be required for the disruption of the interaction between tau and MTs. Additionally, it remains to be determined whether or not all of the phosphorylation sites in tau are involved in the regulation of the binding of tau to MTs since some of these sites may also regulate other yet unknown functions of tau. Indeed, since the phosphorylation and dephosphorylation of these sites appear to be differentially regulated, it is likely that the phosphorylation of some of these sites could subserve a number of different functions. On the basis of the studies reviewed, we concluded (Lee et al. 1991) that the generation of PHF-tau results in whole or in part from the aberrant hyperphosphorylation of normal adult brain tau.

7.5 Implications of the Accumulation of PHF-Tau in the AD Brain

Although several genes have been implicated in the etiology of familial AD (Goate et al. 1991; Strittmatter et al. 1993; Sherrington et al. 1995; Levy-Lahad et al. 1995), suggesting that AD is a polygenic disorder, the etiology of sporadic AD is still unknown. Thus, the search for molecular markers of the pathological state must be continued in the hope that this will provide clues about the cause of AD and enable early diagnosis so that effective therapies can be introduced in a timely manner. In this respect tau is a good candidate marker of AD since its abnormal phosphorylation sites can be used as indicators of the disease at a cellular level. For example, recent studies suggest that cerebrospinal fluid (CSF) tau is a marker of AD. The major goal of these studies was to develop an objective antemortem diagnostic test for AD. Remarkably, the overwhelming concensus that emerges from these five recent papers is that the levels of CSF tau are significantly elevated in AD patients compared to normal elderly controls (Hock et al. 1995; Arai et al. 1995; Jensen et al. 1995; Mori et al. 1995; Vigo-Pelfrey et al. 1995). Initial studies by Hock et al. (1995) demonstrated marked elevations of CSF tau levels in AD patients, and suggested that the amount of tau in the CSF of AD patients correlated with the severity of the dementia. Although elevated CSF levels of tau were also detected in some patients with other acute neurological diseases (e.g., encephalitis, stroke), these diseases are readily distinguished from AD on clinical grounds (Arai et al. 1995; Jensen et al. 1995; Mori et al. 1995; Vigo-Pelfrey et al. 1995). While additional research is needed to confirm and extend these preliminary, but nonetheless provocative studies, the results reported by these different research teams are remarkably concordant. Hence, measurements of CSF tau by ELISA (enyzme-linked immunosorbent assay) or other methods for the quantitation of CSF tau protein may provide an objective antemortem diagnostic test for AD. Indeed, measurement of tau levels (alone or in combination with other potential CSF proteins such as Aβ, fragments of Aβ precursor proteins, kinases, phosphatases, proteases, etc.) in multiple CSF samples from the same patients obtained at periodic intervals spanning all stages of AD may yield powerful new strategies for monitoring the progression of this disease and its response to novel therapeutic agents.

The fact that in the PHF-tau mainly the Ser/Thr residues are phosphorylated points to the deregulation of kinases or phosphatases that might manifest itself in the form of abnormal tau. Despite gaps in our understanding of the detailed pathobiology of the AD neurofibrillary lesions, the available information suggests that the conversion of normal tau into PHF-tau may have deleterious effects on neurons during the progression of AD. As PHF-tau does not bind to the MTs, the accumulation of PHF-tau in the neuronal perikarya above a critical concentration may result in the formation of PHFs. The concomitant depletion of tau below a critical concentration could destabilize microtubules in axons, thereby initiating a cascade of secondary perturbations in the metabolism of other axonal cytoskeletal proteins. The culmination of these events could be axonal degeneration, and the retrograde intraneuronal transport. The notion that accumulations of PHF could physically "block" axonal transport in AD is supported by the recent demonstration of defective axonal transport in a transgenic mouse model of motor neuron disease that was generated by the overexpression of neurofilament proteins and the accumulation of neurofilament aggregates in affected spinal cord neurons (Collard et al. 1995).

The majority of phosphorylated sites in normal brain tau and PHF-tau have been identified by mass spectroscopy and the use of phosphorylation-dependent anti-tau antibodies, thus providing a tool for identifying protein kinases/phosphatase that regulate the phosphorylation state of tau. Thus various candidate enzymes have been described that phosphorylate tau at some of the relevant sites in vitro. Similarly, apolipoprotein E has been shown to bind to recombinant tau in vitro, resulting in the proposal that it may modulate the rate at which tau becomes phosphorylated in AD brain (Strittmatter et al. 1994). However, these studies suffer from the inherent limitation that the in vitro studies are not necessarily representative of the in vivo state. Hence, it would be essential to relate these in vitro findings to the in vivo situation.

Until very recently, studies in the abnormal phosphorylation of PHF-tau focused on the identification of the kinases that might be involved in the generation of PHFs in AD, while little attention was paid to the possible involvement of phosphatase(s) in the pathogenesis of neurofibrillary tangles in AD. Although we cannot specify the precise mechanisms leading to the generation of PHF-tau in AD brain, a decrease in the expression of phosphatases or an increase in the levels of phosphatase inhibitors in the AD brain may also contribute to the generation of PHF-tau, the in vivo deposition of PHFs in neurons, and possibly other AD-associated neuropathological changes. Our results (Mawal-Dewan et al. 1994) suggest that the developing brain may be a useful system in which to study the kinases and phosphatases involved in regulating the dynamic phosphorylation and dephosphorylation of tau. The elucidation of the mechanisms underlying these phenomenon could lead to insights into the pathological events involved in the development of the neurofibrillary pathology in the AD brain.

Acknowledgments. We thank our colleagues in the Department of Pathology and Laboratory Medicine, Neurology and the Penn Alzheimer Center for their many contributions to the work reviewed here which was supported by the grants from the N.I.H. We also thank the families of the patients whom we have studied for their important contributions to this research.

References

Alzheimer A (1907) Über eine eigenartige Erkrankung der Hirnrinde. Allg Z Psychiatr Ihre Grenzgeb 64:146–148

Arai H, Masanori T, Miura M, Higuchi S, Muramatsu T, Machida N, Seki H, Takase S, Clark CM, Lee VMY, Trojanowski JQ, Sasaki H (1995) Tau in cerebrospinal fluid: a potential diagnostic marker of Alzheimer's disease. Ann Neurol 38:649–652

Arriagada PA, Growdon JH, Hedley-White ET, Hyman BT (1992) Neurofibrillary tangles but not senile plaques parallel duration and severity of Alzheimer's disease. Neurology 42:631–639

Baumann K, Mandelkow EM, Biernat J, Piwnica-Worms H, Mandelkow E (1993) Abnormal Alzheimer-like phosphorylation of tau protein by cyclin-dependent kinase cdk2 and cdk5. FEBS Lett 336:417–424

Biernat J, Gustke N, Drewes G, Mandelkow EM, Mandelkow E (1993) Phosphorylation of ser-262 strongly reduces binding of tau to microtubules: distinction from PHF-like immunoreactivity and microtubule binding. Neuron 11:153–163

Braak H, Braak E (1991) Neuropathological stageing of Alzheimer-related changes. Acta Neuropathol (Berl) 82:239–259

Braak H, Braak E, Grundke-Iqual I, Iqbal K (1986) Occurrence of neuropil threads in the senile human brain and in Alzheimer's brain: a third location of paired helical filament outside of neurofibrillary fragments and neuritic plaques. Neurosci Lett 65:351–355

Bramblett GT, Trojanowski JQ, Lee VMY (1992) Regions with abundant neurofibrillary pathology in human brain exhibit a selective reduction in levels of binding-competent tau and the accumulation of abnormal tau isoforms. Lab Invest 66:212–222

Bramblett GT, Goedert M, Jakes R, Merrick SE, Trojanowski JQ, Lee VMY (1993) Abnormal tau phosphorylation at Ser-396 in Alzheimer's disease recapitulates development and contributes to reduced microtubule binding. Neuron 10:1089–1099

Brion JP, Hanger DP, Couck AM, Anderton B (1991) A68 proteins in Alzheimer's disease are composed of several tau isoforms in a phosphorylated state which affects their electrophoretic mobility. Biochem J 279:831–836

Brion JP, Smith C, Couck AM, Gallo JM, Anderton BH (1993) Developmental changes in tau phosphorylation: fetal tau is transiently phosphorylated in a manner similar to paired helical filament tau characteristic of Alzheimer's disease. J Neurochem 61:2071–2080

Butner KA, Kirschner MW (1991) Tau protein binds to microtubules through a flexible array of distributed weak sites. J Cell Biol 115:717–730

Cleveland DW, Hwo SY, Kirshner MW (1977) Physical and chemical properties of purified tau factor and the role of tau in microtubule assembly. J Mol Biol 116:227–247

Collard JP, Cote F, Julien JP (1995) Defective axonal transport in a transgenic mouse model of amyotrophic lateral sclerosis. Nature 375:61–64

Couchie D, Mavilia C, Georgieff IS, Liem RKH, Shelanski ML, Nunez J (1992) Primary structure of high molecular weight tau present in the peripheral nervous system. Proc Natl Acad Sci USA 89:4378–4381

Crowther RA, Olesen OF, Jakes R, Goedert M (1992) The microtubule repeats of tau protein assemble into filament like those found in Alzheimer's disease. FEBS Lett 309:199–202

Dickson DW, Crystal HA, Mattiace, LA, Masur, DM, Blau A D, Davies P, Yen SH, Aronson M (1991) Identification of normal and pathological aging in prospectively studied nondemented elderly humans. Neurobiol Aging 13:179–189

Dreschel DN, Hyman AA, Cobbs MH, Kirschner M (1992) Modulation of the dynamic instability of tubulin assembly by the microtubule-associated protein tau. Mol Biol Cell 3:1147–1154

Drewes G, Lichtenberg-Kraag B, Doring F, Mandelkow EM, Biernat J, Goris J Doree M, Mandelkow E (1992) Mitogen activated protein (MAP) kinase transforms tau protein into an Alzheimer-like state. EMBO J 11:2131–2138

Drewes G, Mandelkow EM, Baumann K, Goris K, Merlevede W, Mandelkow E (1993) Dephosphorylation of tau protein and Alzheimer paired helical filaments by calcineurin and phosphatase-2A. FEBS Lett 336:425–432

Drewes G, Trinczek B, Illenberger S, Biernat J, Schmitt-UlmsG, Meyer HE, Mandelkow EM, Mandelkow E (1995) Microtubule associated protein/microtubule affinity-regulating kinase (p110^{mark}). J Biol Chem 270:7679–7688

Garver TD, Harris KA, Lehman RAW, Trojanowski JQ, Billingsley ML(1994) Tau phosphorylation in human, primate and rat brain: evidence that a pool of tau is highly phosphorylated in vivo and is rapidly dephosphorylated in vitro. J Neurochem 63:2279–2287

Goate A, Chartier-Harlin MC, Mullan M, Brown J, Crawford F, Fidani L, Roques P, Talbot C, Pericak-Vance M, Roses AD, Williamson R, Rossor M, Owen M, Hardy J (1991) Segregation of a missence mutation in the amyloid precursor protein gene with familial Alzheimer's disease. Nature 349:704–706

Goedert M (1993) Tau protein and the neurofibrillary pathology of Alzheimer's disease. Trends Neurosci 16:460–465

Goedert M, Jakes R (1990) Expression of separate isoforms of human tau protein: correlation with the human tau protein in brain and effects on tubulin polymerization. EMBO J 9:4225–4230

Goedert M, Wischik CM, Crowther RA, Walker JE, Klug A (1988) Cloning and sequencing of the cDNA encoding core protein of the paired helical filament of Alzheimer's disease: identification as the microtubule associated protein tau. Proc Natl Acad Sci USA 85:4051–4055

Goedert M, Spillantini MG, Jakes R, Rutherford D, Crowther RA (1989a) Multiple isoforms of human microtubule-associated protein tau: sequences and localization in neurofibrillary tangles of Alzheimer's disease. Neuron 3:519–526

Goedert M, Spillantini MG, Potier MC, Ulrich J, Crowther RA (1989b) Cloning and sequencing of the cDNA encoding an isoform of microtubule-associated protein tau containing four tandem repeats: differential expression of tau protein mRNAs in human brain. EMBO J 8:393–399

Goedert M, Spillantini MG, Crowther RA (1992a) Cloning of a big tau microtubule associated protein characteristic of the peripheral nervous system. Proc Natl Acad Sci USA 89:1983–1987

Goedert M, Spillantini MG, Cairns NJ, Crowther RA (1992b) Tau protein of Alzheimer paired helical filaments: abnormal phosphorylation of all six brain isoforms. Neuron 8:159–168

Goedert M, Cohen ES, Jakes R, Cohen P (1992c) p42 MAP kinase phosphorylation sites in microtubule-associated protein tau are dephosphorylated by protein phosphatase 2A1. Implications for Alzheimer's disease. FEBS Lett 312:95–99

Goedert M, Jakes R, Crowther RA, Six J, Lubke U, Vandermeeren M, Cras P, Trojanowski JQ, Lee VMY (1993) The abnormal phosphorylation of tau proteins at Ser-202 in Alzheimer's disease recapitulates phosphorylation during development. Proc Natl Acad Sci USA 90:5066–5070

Goedert M, Jakes R, Spillantini MG, Crowther RA (1994a) Tau protein and Alzheimer's disease. In: Hyams JS, Lloyd CW (eds) Microtubules. Wiley-Liss, New York, pp 183–200

Goedert M, Jakes R, Crowther RA, Cohen P, Vanmechelen E, Vandermeeren M, Cras P (1994b) Epitope mapping of monoclonal antibodies to the paired helical filaments of Alzheimer's disease: identification of phosphorylation sites in tau protein. Biochem J 301:871–877

Goedert M, Jakes R, Vanmechelen E (1995) Monoclonal antibody AT8 recognizes tau protein phosphorylated at both Ser-202 and Thr-205. Neurosci Lett 189:167–170

Greenberg SG, Davies P (1990) A preparation of Alzheimer paired helical filaments that displays distinct tau proteins by polyacrylamide gel electrophoresis. Proc Natl Acad Sci USA 87:5827–5831

Greenberg SH, Davies P, Schein P, Binder LI (1992) Hydrofloric acid treated PHF-tau proteins display the same biochemical properties as normal tau. J Biol Chem 267:564–569

Hanger DP, Hughes K, Woodgett JR, Brion JP, Anderton BH (1992) Glycogen synthase kinase 3 induces Alzheimer's disease-like phosphorylation of tau: generation of paired helical filaments epitope and neuronal localization of the kinase. Neurosci Lett 147:58–62

Harris KA, Oyler GA, Doolittle GM, Vincent I, Lehman RAW, Kincaid RL, Billingsley ML (1993) Okadaic acid induces hyperphosphorylated forms of tau protein in human brain slices. Ann Neurol 33:77–87

Hasegawa M, Morishima-Kawashima M, Takio K, Suzuki M, Litani K, Ihara Y (1992) Protein sequence and mass spectrometric analysis of tau in the Alzheimer's disease brain. J Biol Chem 267:17047–17054

Hasegawa M, Watanabe A, Takio K, Suzuki M, Arai T, Titani K, Ihara Y (1993) Characterization of two distinct monoclonal antibodies to paired helical filaments: further evidence for fetal type phosphorylation of the tau in paired helical filaments. J Neurochem 60:2068–2077

Hock C, Golomboswki S, Naser W, Mueller-Spahn F (1995) Increased levels of tau in cerebrospinal fluid of patients with Alzheimer's disease – correlation with degree of cognitive impairment. Ann Neurol 183:43–45

Jensen M, Basum H, Lannfelt L (1995) Increased cerebrospinal fluid tau in patients with Alzheimer's disease. Neurosci Lett 186:186–191

Joachim CL, Selkoe DJ (1992) The seminal role of β-amyloid in the pathogenesis of Alzheimer's disease. Alzheimer Dis Assoc Disord 6:7–34

Kanemura K, Takio K, Miura R, Titani K, Ihara Y (1992) Fetal type phosphorylation of the tau in paired helical filaments. J Neurochem 58:1667–1675

Kenessey A, Yen SHC (1993) The extent of phosphorylation of fetal tau is comparable to that of PHF-tau from Alzheimer paired helical filaments. Brain Res 629:40–46

Khatoon S, Gondke-Iqual I, Iqbal K(1992) Brain levels of microtubule-associated protein tau are elevated in Alzheimer's disease: a radioimmunoslotblot assay for nanograms of the protein. J Neurochem 58:1667–1675

Kidd M (1963) Paired helical filaments in electron microscopy of Alzheimer's disease. Nature 197:192–193

Kobayashi S, Ishiguro K, Omori A, Takamatsu M, Arioka M, Imahori K, Uchida T (1993) A cdc-related kinase PSSALRE/cdk5 is homologous with the 30 kD subunit of tau protein kinase II, a proline directed protein kinase associated with microtubule. FEBS Lett 335:171–75

Kosik KS (1992) Alzheimer's disease: a cell biological perspective. Science 256:780–783

Kosik KS, Orecchio LD, Binder L, Trojanowski JQ, Lee VMY, Lee G (1988) Epitopes that span the tau molecule are shared with paired helical filaments. Neuron 1:817–825

Ksiezak-Reding H, Yen SH (1991) Structural stability of paired helical filaments requires microtubule-binding domains of tau: a model for self-association. Neuron 6:717–728

Ksiezak-Reding H, Binder L, Yen SH (1990) Alzheimer's disease proteins (A68) share epitopes with tau but distinct biochemical properties. J Neurosci Res 25:420–430

Ledesma MD, Correas I, Avila J, Diaz-Nido J (1992) Implication of brain cdc2 and MAP kinase in the phosphorylation of tau protein in Alzhiemer's disease. FEBS Lett 308:218–224

Lee VMY, Balin BJ, Otvos L Jr, Trojnowski JQ (1991) A68: a major subunit of paired helical filaments and derivatized forms of normal tau. Science 251:675–678

Lee JHM, Goedert M, Hill WD, Lee VMY, Trojanowski JQ (1993) Tau proteins are abnormally expressed in olfactory epithelium of Alzheimer's disease and developmentally regulated in fetal spinal cord. Exp Neurol 121:1667–1675

Levy-Lahad E, Wasco W, Porkaj P, Romano DM, Oshima J, Petingell WH, Yu C, Jondro PD, Schmidt SD, Wang K, Crowley AC, Fu YH, Guenette SY, Galas D, Nemens E, Wijsman EM, Bird TD, Schellenberg GD, Tanzi RE (1995) Candidate gene for the chomosome 1 familial Alzheimer's disease locus. Science 269:973–977

Linwall G, Cole RD (1984) Phophorylation affects the ability of tau to promote microtubule assembly. J Biol Chem 259:5301–5305

Litersky JM, Johnson GVW, Jakes R, Goedert M, Lee M, Seubert P (1995) Tau protein is phosphorylated by cAMP-dependent protein kinase and calcium/calmodulin-dependent protein kinase II within its microtubule binding domains at Ser-262 and Ser-356. J Biol Chem 270:18917–18922

Liu WK, Ksiezak-Reding H, Yen SH (1991) Abnormal tau proteins from Alzheimer's disease brains: purification and amino acid analysis. J Biol Chem 266:21723 –21727

Lodestone S, Reynolds CH, Latimer D, Davis DR, Anderton BH, Gallo JM, Hanger D, Mulot S, Marquardt B, Stabel S, Woodgett JR, Miller CCJ (1994) Alzheimer disease like phosphorylation of the microtubule-associated protein tau by glycogen synthase kinase-3 in transfected mammalian cells. Curr Opin Biol 4:1077–1086

Mandelkow EM, Drewes G, Biernat J, Gustke N, Van Lint J, Vandenheede JR, Mandelkow E (1992) Glycogen synthase kinase 3 and the Alzheimer like state of microtubule associated protein tau. FEBS Lett 314:315–321

Matsuo ES, Shin RW, Billingsley ML, Van de Voorde A, O'Connor M, Trojanowski JQ, Lee VMY (1994) Biopsy derived adult human brain tau is phosphorylated at many of the same sites as Alzheimer's disease paired helical filament. Neuron 13:989–1002

Mawal-Dewan M, Sen PC, Abdel-Ghany M, Shalloway D, Racker E (1992) Phosphorylation of tau protein by purified p34cdc28 and a related protein kinase from neurofilaments. J Biol Chem 267:19705–19709

Mawal-Dewan M, Henley J, Van de Voorde A, Trojanowski JQ, Lee VMY (1994) The phosphorylation state of tau in the developing rat brain is regulated by phosphoprotein phosphatases. J Biol Chem 49:30981–30987

Mori H, Hosoda K, Matsubara E, Nakamoto T, Furiya Y, Endoh R, Usami M, Shoji M, Maruyama S, Hirai S (1995) Tau in cerebrospinal fluids: establishment of the sandwich ELISA with antibody specific to the repeat sequence in tau. Neurosci Lett 186:181–183

Morishima-Kawashima M, Hasegawa M, Takio K, Suzuki M, Yoshida H, Titani K, Ihara Y (1995) Proline-directed and non-directed phosphorylation of PHF-tau. J Biol Chem 270:823–829

Paudel HK, Lew J, Zenobia A, Wang JH (1993) Brain proline-directed protein kinase phosphorylates tau on sites that are abnormally phosphorylated in tau associated with Alzheimer's paired helical filaments. J Biol Chem 268:23512–23518

Poulter L, Barratt D, Scott CW, Caputo CB. (1993) Location and immunoreactivities of phosphorylation sites on bovine and porcine tau proteins and a PHF-tau fragment. J Biol Chem 268:9636–9644

Seubert P, Mawal-Dewan M, Barbour R, Jakes R, Goedert M, Johnson GVW, Litersky JM, Schenk D, Lieberburg I, Trojanowski JQ, Lee VMY (1995) Detection of phosphorylated Ser-262 in fetal tau, adult tau and paired helical filament tau. J Biol Chem 32:18917–18922

Sherrington R, Rogaev EI, Liang Y, Rogaeva EA, Levesque G, Ikeda M, Chi H, Lin C, Li C, Li G, Holman K, Tsuda T, Mar L, Foncin JF, Bruni AC, Montesi MP, Sorbi S, Painero I, Pinessi L, Nee L, Chumakov I, Pollen D, Brookes A, Sanseau P, Polinsky RJ, Wasco W, Da Silva HAR, Haines JL, Pericak-Vance MA, Tanzi RE, Roses AD, Fraser PE, Rommens JM, St George-Hyslop PH (1995) Cloning of a gene bearing missence mutations in early-onset familial Alzheimer's disease. Nature 375:754–760

Shirazi SK, Wood JG (1993) The protein tyrosine kinase, fyn, in Alzheimer disease pathology. Neurol Rep 4:435–437

Strittmatter WJ, Saunders AM, Schmechel D, Pericak-Vance M, Enghild J, Salvesen GS, Roses AD (1993) Apolipoprotein E: high avidity binding to β-amyloid and increased frequency of type 4 allele in late-onset familial Alzheimer's disease. Proc Natl Acad Sci USA 90:1977–1981

Strittmatter WJ, Saunders AM, Goedert M, Weisgraber KH, Dong LM, Jakes R, Huang DY, Pericak-Vance M, Schmechel D, Roses AD (1994) Isoform-specific interactions of apolipoprotein E with microtubule-associated protein tau: implications for Alzheimer's disease. Proc Natl Acad Sci USA 91:11183–11186

Tagliavani F, Giaccone G, Prelli F, Verga L, Porro M, Trojanowski JQ, Farlow MR, Frangione B, Ghetti B, Bugiani O (1993) A68 is a component of paired helical filaments of Gerstmann-Straessler-Scheinker disease, Indiana kindred. Brain Res 616:325–328

Tohyama T, Lee VMY, Rorke LB, Trojanowski JQ (1991) Molecular milestones that signal axonal maturation and the commitment of human spinal cord precursor cells to the neuronal or glial phenotype in development. J Comp Neurol 310:285–299

Trojanowski JQ, Schmidt ML, Otvos L Jr, Arai H, Hill WD, Lee MY (1990) Vulnerability of the neuronal cytoskeleton in Alzheimer's disease: widespread involvement of three major filament systems. Annu Rev Gerontol Geriatr 10:167–182

Trojanowski JQ, Schmidt ML, Shin RW, Bramblett GT, Rao D, Lee VMY (1993a) Altered tau and neurofilament proteins in neurodegenerative diseases: diagnostic implications for Alzheimer's disease and Lewy body dementias. Brain Pathol 3:45–54

Trojanowski JQ, Schmidt ML, Shin RW, Bramblett GT, Goedert M, Lee VMY (1993b) PHFτ (A68): from pathological marker to potential mediator of neuronal dysfunction and degeneration in Alzheimer's disease. Clin Neurosci 1:184–191

Trojanowski JQ, Mawal-Dewan M, Schmidt ML, Martin J, Lee VMY (1993c) Localization of the mitogen activated protein kinase ERK2 in Alzheimer's disease neurofibrillary tangles and senile plaque neurite. Brain Res 618:333–337

Vigo-Pelfrey C, Suebert P, Barbour R, Blomquist C, Lee M, Lee D, Coria F, Chang L, Miller B, Lieberburg I, Shenk D (1995) Elevation of microtubule-associated protein tau in cerebrospinal fluid of patients with Alzheimer's disease. Neurology 45:788–793

Vincent IJ, Davies P (1990) Phosphorylation characteristics of the A68 protein in Alzheimer's disease. Brain Res 531:127–135

Vulliet R, Halloran SM, Braun RK, Smith AJ, Lee G (1992) Proline directed phosphorylation of human tau protein. J Biol Chem 267:22570–22574

Watanabe A, Hasegawa M, Suzuki M, Takio K, Morishima-Kawashima M, Titani K, Arai T, Kosik KS, Ihara Y (1993) Invivo phosphorylation sites in fetal and adult rat tau. J Biol Chem 268:25712–25716

Wille H, Drewes G, Biernat J, Mandelkow EM, Mandelkow E (1992) Alzheimer-like paired helical filaments and anti-parallel dimers formed from microtubule associated tau in vitro. J Cell Biol 118:573–584

Wischik CM, Novak M, Thogersen HC, Edwards PC, Runswick MJ, Jakes R, Walker J, Milstein C, Roth A, Klug A (1988) Isolation of a fragment of tau derived from the core of the paired helical filament of Alzheimer's disease. Proc Natl Acad Sci USA 85:4506–4510

Yachnis AT, Rorke LB, Lee VMY, Trojanowski JQ (1993) Expression of neuronal and glial polypeptides during histogenesis of the human cerebellar cortex including observations on the dentate nucleus. J Comp Neurol 334:356–369

Yoshida H, Ihara Y (1993) Tau in paired helical filaments is functionally distinct from fetal tau. J Neurochem 61:1183–1186

8 Cytokines in Alzheimer's and Other Neurodegenerative Diseases

N.J. Rothwell

8.1 Introduction to Cytokines

The dramatic surge of interest in the neurobiology of cytokines seen over the past few years has been due in large part to their implied roles in neurological disease. However, these fascinating molecules were originally identified as messengers of the immune system, and mediators of tissue inflammation, cell differentiation and growth, and the acute phase responses to disease and injury.

The umbrella term "cytokine" now describes a large group of polypeptides with diverse functions, and complex nomenclature, including interleukins, tumour necrosis factors, interferons and growth factors/neurotrophins (NT). In many cases, their naming and grouping as families relate more to historical discovery than to specific functions. There are, however, some general features of cytokines. As polypeptides, of about 8–25 kDa in size, cytokines are very rapidly and often transiently produced by many cell types. Constitutive expression is generally low, and synthesis is induced by a variety of inflammatory or immune stimuli. Some exceptions are those cytokines implicated in growth and differentiation, which may be produced in significant quantities during development. A number of cytokines (such as interleukin-1, IL-1) seem to be ubiquitous, since they are produced by virtually all cell types, while others are relatively cell-specific.

In general, cytokines act in a predominantly autocrine or paracrine manner on cells in close proximity to their site of synthesis. Again some exceptions can be found, perhaps most notably interleukin-6, which appears in circulation during many forms of systemic injury or disease (plasma concentrations can increase by over three orders of magnitude), and shows many features of a classical hormone.

Pleiotropism, overlapping actions and remarkable potency are general features of cytokines, which are important in considering their potential use in therapies for neurological disorders. Most cytokines have diverse actions on many cell types, are active in the femtomolar or picomolar range, and can elicit responses in cells which apparently possess as few as ten receptors. Thus, while apparently beneficial effects of a specific cytokine on neuronal growth or survival may offer an attractive opportunity for therapy, numerous and sometimes unexpected side effects may result. Similarly, inhibition of specific cytokines may elicit unpredictable responses. The cytokine network is highly complex; very different cytokines often share common actions and can induce themselves and other related molecules, making the distinction of their specific actions or roles a daunting task.

Another interesting feature of cytokines is the existence of numerous inhibitory or regulatory factors, which is perhaps not surprising in view of their diversity of action. While some cytokines act synergistically (e.g. IL-1 and tumour necrosis factor, TNF-α), others are antagonistic – a fact which may underlie the apparent benefit of interferon-β (Betaferon) in multiple sclerosis, since it is believed to act by suppressing the synthesis and actions of interferon-β (Panitch 1991). Several "anti-inflammatory cytokines" including IL-4, IL-10 and transforming growth factor (TGF)-β, have been identified. As discussed later, some cytokines are clearly neuroprotective, while others apparently mediate neuronal death. One of the most fascinating cytokines is IL-1 receptor antagonist (IL-1ra), a selective and highly competitive antagonist at

IL-1 receptors, with no apparent agonist activities (Dinarello and Thompson 1991). IL-1ra is a valuable tool for experimental studies, but also appears to function as a biological inhibitor of IL-1 action in the brain as well as the periphery. Cytokine actions may also be influenced by release of soluble receptors which bind extracellular molecules and generally inhibit their actions, although in some cases such receptors can enhance cytokine activity through prolongation of biological half-life.

Research on the neurobiology of cytokines has developed rapidly over the last few years (see Hopkins and Rothwell 1995). However, the first biological actions ascribed to a cytokine were on fever when, in 1948, endogenous pyrogen was found to be a circulating protein, later identified as IL-1. Fever is just one of many actions of cytokines in the brain, and many of these molecules have now been identified in the CNS.

8.2 Expression of Cytokines in the Brain

Because of their size, cytokines produced outside the CNS cannot readily enter the brain by passive diffusion. However, active transport systems have been identified for some cytokines (e.g. IL-1, IL-1ra, TNF-α and IL-6) which probably allow relatively small amounts to enter the brain (Banks et al. 1989; Gutierrez et al. 1993, 1994; Luheshi et al. 1994). In neurological disease or brain injury, breakdown of the blood–brain barrier or invasion of peripheral immune cells (a rich source of cytokines) may contribute to brain cytokines. However, it is now clear that many cytokines can be synthesised in the brain by resident cells (neurones, glia, endothelial cells), and that this expression is markedly increased during any form of damage to the brain (see Hopkins and Rothwell 1995).

As in peripheral tissues, cytokine expression is low (often barely detectable) in adult healthy brain, although the remarkable potency of these molecules means that possible roles in normal physiological functions cannot be fully excluded (see below). In parallel with the rapid growth of this field, numerous publications have reported increases in cytokine expression (particularly the pro-inflammatory cytokines such as IL-1, IL-6, TNF-α) in the brains or cerebrospinal fluid of patients with virtually all neurological diseases, including Alzheimer's disease (see below). Such reports imply diverse roles for these molecules in neurodegeneration, but must be interpreted with caution. Cytokines are usually induced by inflammation and cell damage, thus their expression in neurodegenerative disease may be a result rather than a causal factor in the condition. However, the very early detection of IL-1 in patients with head injuries (e.g. McClain et al. 1987), and the finding that this cytokine is overexpressed in the brains of patients suffering from Down syndrome prior to gross degeneration is supportive of direct involvement (Griffin et al. 1989).

More substantial evidence for cytokine involvement in neuronal damage and death derives from experimental studies in animals. The time course of expression of cytokines in rodents after stroke or head injury again supports some direct involvement in subsequent pathological events. After transient (Minami et al. 1992; Yabuuchi et al. 1994) or focal cerebal ischaemia (Liu et al. 1993; Buttini et al. 1994), IL-1β mRNA is elevated within 30 min, and bioactive and immunoreactive protein is present within hours after stroke or experimental brain injury. Microglia appear to be a major cell source of pro-inflammatory cytokines such as IL-1 (Lee et al. 1993), although immunohistochemical localisation and studies on isolated brain cells in vitro indicate that neurones, glia and endothelial cells can produce cytokines (see Hopkins and Rothwell 1995).

These observations are consistent with a direct involvement of cytokines in acute, and possibly chronic neurodegenerative diseases. However, their roles and actions may be complex, and the establishment of cytokine expression in such diseases does not reveal whether they act to inhibit, promote or cause disease progression.

8.3 Elucidating Actions of Cytokines in the Brain

The number and diversity of cytokine effects on the brain are now vast (see Rothwell and Hopkins 1995 for review) and outside the scope of this chapter. Instead, a brief appraisal of the approaches which have been adopted and the primary actions of cytokines on neuronal survival and death may be useful.

The general interest in the neurobiology of cytokines and the wide availability of recombinant cytokines have provoked extensive research, but also havedrawbacks. Numerous reports now exist (over 3000 publications on various aspects of cytokines and the nervous system, most in the past 2 years) on effects of applying cytokines to brain cells in vitro or injection into the brain of experimental animals. While much valuable information has derived from these approaches, they do not necessarily reveal biological functions of a cytokine or its specific role in disease, and have often yielded conflicting data.

An excellent example of this is found in studies on neuroprotective or neurotrophic actions of cytokines. IL-1 and TNF-α have been implicated in the processes of neurodegeneration in vivo, and in the case of IL-1 there is now substantial evidence that it mediates neuronal death (see below). However, several studies, including some from our own laboratory, have demonstrated clear and potent *neuroprotective* effects of IL-1 and TNF-α in primary cultured neurones in vitro (Araujo 1992; Strijbos and Rothwell 1995). This example illustrates the complexity of cytokine action and the problems associated with experimental approaches. Numerous differences exist between the culture systems employed and in vivo experiments, including the maturity of nerves, synaptic connections, local environment and, perhaps most importantly, the low glial content in most primary neuronal cultures. The problem of establishing appropriate in vitro systems to study cytokines is particularly relevant to Alzheimer's disease for which in vivo experimental models are, at best, highly limited.

On the basis of in vivo studies, neuroprotective or neurotrophic actions of growth factors and neurotrophins (e.g. NGF, GDNF, BDNF, CNTF, FGF, TGF-β) have been clearly demonstrated, in some cases on specific neuronal types. These observations provide considerable hope and expectations for novel therapeutic strategies in Alzheimer's disease (e.g. see Carswell 1993) which will not be reviewed here. In contrast, pro-inflammatory cytokines, particularly IL-1 and TNF-α, exacerbate many forms of experimental brain injury, cause glial activation, blood–brain barrier disruption and release of numerous potentially neurotoxic or inflammatory molecules (see Rothwell and Relton 1993; Rothwell and Hopkins 1995). However, although such in vivo studies have greater clinical relevance than experiments in vitro, observations on effects of exogenously administered cytokines do not provide conclusive evidence of the role of the *endogenous* molecule, or of the specificity or selectivity of action. The latter point may be particularly relevant to considerations of the chronic use of very potent cytokines with diverse actions. The most direct approach to elucidating the biological role of cytokines in neurodegenerative disease is to block their actions in experimental models in vivo.

8.4 Involvement of Cytokines in Neurodegeneration

Various strategies are now available to inhibit cytokine action in vivo, including passive immunoneutralisation of the cytokine or its receptor, administration of soluble receptors or binding proteins, or inhibitors of synthesis (usually by inhibition of converting enzymes, now being developed for IL-1 and TNF-α). One of the most valuable and widely used tools to inhibit a cytokine is IL-1ra, the naturally occurring antagonist of IL-1. We have used this molecule and other approaches to investigate the role of IL-1 in experimentally induced neurodegeneration.

Several years ago, we reported that intracerebroventricular (icv) injection of IL-1ra markedly (over 60%) inhibits neuronal damage caused by focal cerebral ischaemia (middle cerebral artery occlusion, MCAO) in the rat (Relton and Rothwell 1992). From subsequent studies, it is apparent that these effects are not related to changes in body temperature or cardiovascular responses, and that treatment with IL-1ra is effective when delayed up to 30 min after MCAO, and is sustained for at least 7 days (Loddick and Rothwell 1996). The effects of IL-1ra are comparable to, or greater than, those of the *N*-methyl-D-asparate (NMDA) receptor antagonist MK801 and, unlike many other agents, IL-1ra protects striatal as well as cortical brain tissue (Loddick and Rothwell 1996). Relton et al. (1993) have subsequently shown that damage caused by MCAO is also inhibited by IL-1ra injected systemically, and dramatic protective effects have been observed with systemic administration of IL-1ra in neonatal hypoxia in the rat (Martin et al. 1994) and focal ischaemia caused by neurofilament occlusion (Garcia et al. 1995). Betz et al. (1995) have also shown that gene transfer of IL-1ra in vivo, by injection of an adenovirus in which the IL-1ra gene has been inserted, dramatically inhibits MCAO damage in the rat. It is likely (though no published data yet exist) that IL-1ra will be similarly effective against reversible ischaemia, since Kogure's group (Yamasaki et al. 1995) has shown that icv injection of a neutralising anti-IL-1β antibody in the rat dramatically reduces transient ischaemic damage, oedema and neutrophil invasion, and this effect was sustained even when the antibody was injected 4 h after ischaemia.

IL-1ra also markedly inhibits neuronal damage caused by experimental brain injury. Central injection of IL-1ra causes sustained (at least 7 days) inhibition of neurodegeneration caused by lateral fluid percussion injury in the rat (Toulmond and Rothwell 1995). Our recent data (Toulmond and Rothwell, unpublished) further indicate that protection can be achieved by administration of IL-1ra at least 4 h after brain injury.

These findings indicate strongly that endogenous brain IL-1 (probably IL-1β) mediates acute experimental neurodegeneration, and that IL-1ra is a potent neuroprotective agent. Indeed, we now have evidence that *endogenous* IL-1ra acts to limit neuronal damage since it is induced rapidly after ischaemic or excitotoxic brain damage, and neutralisation of brain IL-1ra markedly enhances infarct size (Toulmond, Loddick and Rothwell, unpublished data). Thus, IL-1ra and other selective inhibitors of IL-1 synthesis or action may have therapeutic value.

Considerably less evidence is available for the role of other endogenous cytokines in neurodegeneration, Yamasaki et al. (1994) showed that blocking brain IL-8 action by passive immunoneutralisation inhibits neuronal damage and oedema caused by reversible ischaemia in the rat. Our preliminary data (unpublished) also imply a role for TNF-α in ischaemic brain damage, although IL-6 appears to be neuroprotective. The latter observation contrasts with the report of Campbell et al. (1993) who observed dramatic neurodegeneration and brain inflammation on transgenic mice overexpressing IL-6. This apparent discrepancy may reflect acute versus chronic effects of IL-6, the importance of the magnitude of brain IL-6 concentrations, or specific involvement of IL-6 in brain development, which may be important in transgenic mice.

A similarly complex situation appears to exist for TGF-β1. Injection of recombinant TGF-β1 inhibits ischaemic brain damage (Gross et al. 1993; Henrich-Noack et al. 1994). However, TGF-β1 also inhibits wound healing after brain injury, while application of anti-TGF-β antibody promotes more rapid healing (Logan et al. 1994).

8.5 Potential Mechanisms of Action in Neurodegeneration

Very recent data have suggested that neuroprotective effects of growth factors are due to inhibition of increases in neuronal calcium (see Cheng and Mattson 1994; Cheng et al. 1991, 1993; Mattson 1990; Mattson and Cheng 1993).

In contrast, very little is known about the mechanisms of neurotoxicity of cytokines such as IL-1. IL-1ra dramatically inhibits ischaemic and traumatic damage in vivo, implying that actions of IL-1 are critical in neuronal cell death. However, IL-1 itself, when applied to neuronal cultures (Strijbos and Rothwell 1995), or when infused into the brains of normal rats (Lawrence and Rothwell 1994; Lawrence et al. 1995), does not cause overt neuronal death, at least at concentrations which seem comparable to, or greater than, those observed after neuronal damage. Thus, it is possible that either sustained release of IL-1 is required to cause neurodegeneration, that IL-1 is toxic only to already compromised (e.g. ischaemic) neurones, or that some other cofactor(s) present after damage is involved.

Central (icv) injection of quite low doses of IL-1β markedly exacerbate focal (MCAO, Loddick and Rothwell 1996) or global ischaemic (Yamasaki et al. 1995) or traumatic damage (Toulmond and Rothwell, unpublished data). These effects, particularly on global ischaemia, may be related to hyperthermia, which can worsen ischaemic brain damage directly (Ginsberg et al. 1992). MCAO itself does not affect core body temperature in the rat, but co-administration of IL-1β induces a marked fever (Loddick and Rothwell 1995). While this may contribute to the massive (~ 100%) exacerbation of damage caused by IL-1, it is unlikely to be the sole cause. IL-6 is also a potent pyrogen which causes marked fever in normal or ischaemic rats but, unlike IL-1, *inhibits* rather than enhances MCAO damage (Le Feuvre, Loddick and Rothwell, unpublished data).

We have found that IL-1β does not exacerbate local neuronal damage caused by infusion of an NMDA receptor antagonist into the rat striatum, but has surprising effects on α-amino-3-hydroxy-5-methyl-4-isoxazolepropionate (AMPA) receptor-induced damage. Striatal infusion of S-AMPA causes extensive local damage, and co-infusion of IL-1β, which itself is not toxic, does not affect neurodegeneration in the striatum, but causes extensive neuronal death in the cortex (Lawrence et al. 1995). This indicates some specific interaction between striatal IL-1 and AMPA receptors which activates pathways leading to secondary cortical loss.

In view of the central role of glutamate in ischaemic and traumatic brain injury, it is likely that effects of IL-1ra and IL-1 relate in some way to release or actions of glutamate. We have failed to observe any effects of either IL-1 or IL-1ra on release of glutamate from rat brain slices (Allan and Rothwell, unpublished data), though actions in vivo, particularly in hypoxic brain tissue, cannot be excluded. Additional data suggest that IL-1ra acts at a point beyond glutamate release and activation of excitatory amino acid (EAA) receptors. Co-infusion of IL-1ra markedly inhibits neuronal damage caused by striatal infusion of selective agonists at either NMDA or AMPA receptors (Lawrence and Rothwell 1994; Lawrence et al. 1995). These observations have several implications. Firstly, they suggest a broader involvement of IL-1 (and therapeutic effects of IL-1ra) in different forms of neurodegeneration, since NMDA and AMPA receptor overactivation has been implicated in different forms of neurodegeneration. Secondly, they may provide some clues about mechanisms of action, since NMDA and AMPA receptor activation causes neuronal death through different pathways.

Unfortunately, little advance has been made in understanding neuroprotective effects of IL-1ra from studies on neuronal cultures since IL-1ra does not protect isolated neurones from excitotoxic damage (Strijbos and Rothwell 1995) and, as described above, IL-1 is not toxic to neurones over periods of up to 24 h (Strijbos and Rothwell 1995). Chronic exposure (72 h) of neurones to high concentrations of IL-1 does cause some toxicity, which is not inhibited by antagonists at any of the glutamate receptors, by calcium channel blockers or by a nitric oxide synthase inhibitor (Strijbos and Rothwell 1995).

Several studies have now reported that IL-1 and other cytokines cause glia (particularly microglia) to release toxic substances including glutamate, nitric oxide and other unidentified molecules (Giulian 1993; Giulian and Vaca 1993; Piani et al. 1991). Thus, the presence of microglia and other cell types may be fundamental to actions of IL-1 and IL-1ra. A further example of this is seen in endothelial cells which can produce cytokines (Fabry et al. 1993; Joseph et al. 1993) and express adhesion molecules. Adhesion molecules such as ICAM-1 and V-CAM mediate invasion of immune cells such as lymphocytes and neutrophils into the brain, which themselves may be involved in damage caused by cerebral ischaemia (Clark et al. 1991).

Several other molecules and pathways have been implicated in IL-1 action and neurodegeneration. For example, many cytokines induce release of arachidonic acid which could cause neuronal damage directly or indirectly, through its metabolites such as prostanoids and PAF (platelet-activating factor; see Rothwell and Relton 1993). IL-1 also induces synthesis of the peptide corticotrophin-releasing factor (CRF), which mediates many actions of this cytokine in the CNS including fever and pituitary adrenal activation (Rivier 1993). Inhibition of CRF action by central injection of a CRF receptor antagonist greatly reduces damage caused by global or focal ischaemia or activation of NMDA receptors (Lyons et al. 1991; Strijbos et al. 1994), though the direct relation of these observations to IL-1 release is not known.

Particularly relevant to the present discussion is the possible relationship between cytokines and β-amyloid toxicity. Cytokines (e.g. IL-1 and IL-6) stimulate β-APP synthesis, and TNF-α and interferon-γ synergise with amyloid fragments to cause neuronal death probably through effects on microglia to cause release of nitric oxide (Meda et al. 1995). These actions may be relevant to acute as well as chronic neurodegenerative disease.

The quantitative importance of apoptosis in acute or chronic neurodegeneration remains uncertain, though numerous reports have provided evidence that apoptosis does occur in response to experimentally induced neurodegeneration (see Margolis et al. 1994; Oppenheim 1991). Specific interest has focussed on the possible role of ICE (interleukin-1β converting enzyme) in apoptosis since its homology with ced-3 (an apoptotic gene in *Caenorhabditis elegans*) was identified (Yuan et al. 1993). ICE is a member of a growing family of cysteine proteases, most of which can cause apoptosis when overexpressed. However, it now seems that ICE-1, which cleaves mature IL-1β from its precursor, is not a primary mediator of apoptosis (Nicholson et al. 1995). Nevertheless, ICE may be involved in neuronal death since it is essential for IL-1β release.

The current picture of the mechanisms of cytokine action in neurodegeneration is complex. This is perhaps not surprising given the general complexity of both the cytokine network and the mechanisms of neurodegeneration, but the rapid expansion of research in this field is likely to provide some answers within the next few years.

8.6 Evidence for a Role of Cytokines in Alzheimer's Disease

Some evidence supports the hypothesis that cytokines (particularly IL-1) participate in the development and/or progression of Alzheimer's disease:

1. Overexpression in Alzheimer' disease and Down syndrome.
2. IL-1 and possibly other cytokines can elicit most of the responses which characterise Alzheimer's disease.
3. IL-1 is involved in acute neurodegeneration, which may share common mechanisms with those involved in Alzheimer's disease.
4. IL-1 is involved in damage caused by head injury, which is a significant risk factor associated with Alzheimer's disease.
5. Increasing evidence is emerging for inflammatory responses in the brains of patients with Alzheimer's disease and preliminary studies suggesting benefits of anti-inflammatory drugs.
6. Some data suggest general alterations in immune function in patients with Alzheimer's disease including changes in circulating cytokines or their production by peripheral immune cells.

A number of studies have reported increased expression of cytokines, particularly IL-1, in the brains of patients with Alzheimer's disease compared to age-matched cells. These measurements, usually based on immunocytochemistry, reveal increased amounts of IL-1 and IL-1β (e.g. Cacabelos et al. 1991, 1994; Griffin et al. 1989, 1995). The greatest IL-1 immunoreactivity has been observed in diffuse amyloid plaques with a profusion of β-APP positive dystrophic plaques (Griffin et al. 1995).

IL-1 expression in Alzheimer's disease appears to be largely microglial in origin and is found particularly in brain regions associated with Alzheimer-type pathology, i.e. frontal, parietal and tempral cortex, hippocampus and thalamus, with little of no increase in expression in the cerebellum (Cacabelos et al. 1994). Other cytokines, most notably IL-6, are also reportedly increased in brains of Alzheimer patients (Bauer et al. 1994). In contrast to these studies, which used immunocytochemistry, Wood et al. (1993) reported no difference in mature IL-1β or IL-1ra, detected by ELISA (enzyme-linked immunosorbent assay), in the brains of Alzheimer patients, but observed significantly raised IL-6 concentrations; however, this difference may reflect methodological discrepancies.

The appearance of cytokines in circulation or their production by polymorphonuclear cells is also reportedly elevated in Alzheimer patients. Huberman et al. (1994) observed that mononuclear cells from patients with Alzheimer's disease showed greater IL-2 and interferon (IFN)-γ secretion than those from controls, but less production of IL-1β. In conrast, Singh (1994) described increased production of IL-1, IL-2 and IL-6 by cells from demented patients, while Pirtilla et al. (1994) found that IL-1β was not elevated in cerebrospinal fluid or serum of Alzheimer patients. Fillit et al. (1991) observed increased circulating TNF-α in Alzheimer's disease.

These apparent anomalies are unsurprising in view of the fact that many cytokines (and particularly IL-1) are released into circulation infrequently, transiently and usually in low quantities. These molecules are generally produced and act locally within tissues, though reported changes in cytokine production by immune cells may reflect a fundamental change in sensitivity to stimuli which elicit cytokine production.

The apparent changes in brain cytokine expression in Alzheimer's disease also need to be treated with some caution. Neuronal damage, glial activation and inflammation are common features of Alzheimer's disease, which may *cause* cytokine expression in the brain. Such expression may therefore be secondary to the disease, but could nevertheless exacerbate progression. A second, potentially important factor, rarely considered, is the possibility that demented patients have a higher risk of infection (which may not be readily detected). Systemic infection induces brain cytokine expression (see Hopkins and Rothwell 1995), and could therefore contribute to the findings in Alzheimer patients.

Notwithstanding these caveats about cytokine overexpression and a causal role for these molecules in disease initiation and/or progression, the data described above are consistent with the proposal that cytokines are involved, at least in some way in Alzheimer's disease. Further evidence for this hypothesis derives from the fact that cytokines (of which IL-1 is the most studied) can induce responses in vivo or in vitro which characterise Alzheimer pathology. Of these responses, induction of β-APP, and probably amyloid itself (Buxbaum et al. 1994), has been observed in a number of cells or cells lines in culture, including neurones and glia (Buxbaum et al. 1994; Forloni et al. 1992; Ohyagi and

Tabira 1993; Dash and Moore 1995; Vasilakos et al. 1994). This effect is consistent with the fact that β-APP is considered an acute phase protein. The APP gene contains cytokine response elements, and β-APP mRNA expression is also induced in cultured mouse neuronal or glial cells by IL-2, IL-3, IL-6, NGF, PGF or GM CSF, but surprisingly not by TNF-α (Ohyagi and Tabira 1993), though the bioactivity and species specificity of TNF-α may reflect this negative result.

Buxbaum et al. (1992, 1994) reported that IL-1 stimulation of APP and amyloid production is regulated by calcium, independently of protein kinase C. Dash and Moore (1995) found that IL-1β stimulation of β-APP processing and secretion in PC12 cells is inhibited by indomethacin or NDGA, implicating the cyclooxygenase and lipoxygenase pathways respectively. Yang et al. (1993) described inhibitory effects of IL-1β on APP gene expression in human glioma cell lines, but these cell lines express much lower levels of APP than normal brain and may also have altered cytokine sensitivity.

Overexpression of β-APP in vivo has been observed in response to various forms of brain damage in animals (see Royston et al. 1992) and early after brain trauma in humans. Furthermore, brain trauma has been linked to Alzheimer's disease (Graves et al. 1990; Mayeux et al. 1993; Royston et al. 1992). Few, if any, published studies have described actions of cytokines on APP processing, and future studies to test effects of cytokines or brain insults in transgenic mice with mutated forms or overexpression of β-APP may provide fruitful information.

Many of the pathological changes observed in the brains of Alzheimer patients point towards an inflammatory or altered immune response in the brain, including glial activation, expression of acute phase proteins such as α2 macroglobulin, α1-antichymotrypsin, complement, prostaglandins, MHC class II antigens and adhesion molecules (see Bauer et al. 1991; McGeer et al. 1987, 1989; Vandenabeele and Fiers 1991). All of these responses can be elicited by pro-inflammatory cytokines in experimental animals or cultured cells in vitro.

A further area of interest is cytokine modulation or involvement in β-amyloid neurotoxicity. β-Amyloid and IFN-γ act synergistically to cause neuronal death possibly via TNF-α release (Meda et al. 1995, see above), and amyloid toxicity is associated with release of IL-1, IL-6 and TNF-α, and can be inhibited by TGF-β in human cultured brain cells (Chao et al. 1994).

Inhibition of IL-1 action by IL-1ra attenuates β-APP overexpression induced by brain damage (cerebral ischaemia) in the rat (Royston et al. 1992), and potent effects of cytokine modification on brain inflammatory responses in multiple sclerosis have been reported. For example, inhibition of IL-1 or TNF-α action reduces the clinical symptoms of EAE (experimental allergic encephalomyelitis) in animal models of multiple sclerosis, and interferon β, which shows benefit in patients with this disease, probably acts through inhibition of pro-inflammatory cytokines such as (IL-1, TNF-α, IFN-γ, Panitch 1991). Thus, cytokines could also mediate inflammatory responses in Alzheimer's disease and the reduced disease risk associated with regular use of anti-inflammatory drugs, thus the improvements seen in patients treated with steroidal anti-inflammatory drugs in early trials suggest that inhibiting cytokine action is likely to be beneficial in Alzheimer's disease.

Head injury imposes a significant risk of getting Alzheimer's disease, which may be related to cytokine induction after brain trauma (see Graves et al. 1990; Mayeux et al. 1993; Royston et al. 1992). Thus, it has been postulated that after the acute damage associated with marked overexpression of cytokines (see earlier) persistent but more modest cytokine synthesis and action may lead to chronic pathological changes associated with Alzheimer's disease.

The field of neuroimmunology and the study of the potential involvement of cytokines and inflammation in Alzheimer's disease and related dementias are in their infancy. Results obtained to date, which are likely to be "the tip of the iceberg" are exciting and promising. Such excitement needs to be tempered with caution since it is now obvious that Alzheimer's disease and related dementias are of multifactual origin and complex. Nevertheless, recent studies imply a fundamental role of cytokines in Alzheimer's disease, and further study of this fascinating topic is likely to contribute significantly to our understanding of this important disease and may lead to novel and effective therapies.

References

Araujo DM (1992) Contrasting effects of specific lymphokines on the survival of hippocampal neurones in culture. In: Meyer E (ed) Treatment of dementias. Plenum, New York, pp 113–122

Banks WA, Kastin AJ, Durham DA (1989) Bidirectional transport of interleukin-1 alpha across the blood–brain barrier. Brain Res Bull 23:433–437

Bauer J, Strauss S, Schreiter-Gasser U, Ganter U, Schlegel P, Witt I, Yolk B, Berger M (1991) Interleukin-6 and α2-macroglobulin indicate an acute-phase state in Alzheimer's disease cortices. FEBS Lett 285:111–114

Betz AL, Yang GY, Davidson B (1995) Attenuation of stroke size in rats using an adenoviral vector to induce overexpression of interleukin-1 receptor antagonist in brain. J Cerebr Blood Flow Metab 15:547–551

Buttini M, Sauter A, Boddeke HW (1994) Induction of interleukin-1 beta mRNA after focal ischaemia in the rat. Brain Res 23:126–134

Buxbaum JD, Oishi M, Chen FH, Pinkas-Kramarski R, Jaffe EA, Gandy SE, Greengard P (1992) Cholinergic agonists and IL-1 regulate processing and secretion of the Alzheimer βA4 amyloid precursor. Proc Natl Acad Sci USA 89:10075–10078

Buxbaum JD, Ruefli AA, Parker CA, Cypress AM, Greengard P (1994) Calcium regulates processing of the Alzheimer amyloid protein precursor in a protein kinase C-independent manner. Proc Natl Acad Sci USA 91:4489–4493

Cacabelos R, Barquero M, Garcia P, Alvarez XA, Varela E (1991) Cerebrospinal fluid interleukin 1β (IL-1β) in Alzheimer's disease and neurological disorders. Methods Find Exp Clin Pharmacol 13:455–458

Cacabelos R, Alvarez XA, Femandez-Novoa, L Franco A, Mangues R, Pellicer A, Nishimura T (1994) Brain interleukin-1 on Alzheimer's disease and vascular dementia. Methods Find Exp Clin Pharmacol 16:141–151

Campbell EL, Abraham CR, Masliah E, Kemper P, Inglis JD, Oldstone MBA, Mucke L (1993) Neurologic disease induced in transgenic mice by cerebral overexpression of interleukin-6. Proc Natl Acad Sci USA 90:10061–10065

Carswell S (1993) The potential for treating neurodegenerative disorders with NGF-inducing compounds. Exp Neurol 124:36–42

Chao CC, Hu S, Kravitz FH, Tsang M, Anderson WR, Peterson PK (1994) Transforming growth factor-beta protects human neurons against beta-amyloid-induced injury. Mol Chem Neuropathol 23:159–178

Cheng B, Mattson MP (1991) NGF and bFGF protect rat hippocampal and human cortical neurons against hypoglycemic damage by stabilising calcium homeostasis. Neuron 7:1031–1041

Cheng B, McMahom DG, Mattson MP (1993) Modulation of calcium current, intracellular calcium levels and cell survival by glucose deprivation and growth factors in hippocampal neurons. Brain Res 607:275–286

Cheng B, Christakos S, Mattson MP (1994) Tumor necrosis factors protect neurons against metabolic-excitotoxic insults and promote maintenance of calcium homeostatis. Neuron 12:139–153

Clark WM, Madden KP, Rothlein R, Zivin JA (1991) Reduction of central nervous system ischemic injury by monoclonal antibody to intercellular adhesion molecule. J Neurosurg 75:623

Dash PK, Moore AN (1995) Enhanced processing of APP induced by IL-1 beta can be reduced by indomethacin and nordihydroguaiaretic acid. Biochem Biophys Res Commun 208:542–548

Dinarello CA, Thompson RC (1991) Blocking IL-1: interleukin 1 receptor antagonist in vivo and in vitro. Immunol Today 12:404–410

Fabry Z, Fitzsimmons KM, Herlein JA, Moninger TO, Dobbs MB, Hart MN (1993) Production of the cytokines interleukin 1 and 6 by murine brain microvessel endothelium and smooth muscle pericytes. J Neuroimmunol 47:23–34

Fillit H, Ding W, Buee L, Kalman J, Altstiel L, Lawlor B, Wolf-Klein G (1991) Elevated circulating tumor necrosis factor levels in Alzheimer's disease. Neurosci Lett 129:318–320

Forloni G, Demicheli F, Giorgi S, Bendotti C, Angeretti N (1992) Expression of amyloid precursor protein mRNAs in endothelial, neuronal and glial cells: modutation by interleukin-1. Mol Brain Res 16:128–134

Garcia JH, Liu KF, Relton JK (1995) Interleukin-1 receptor antagonist decreases the number of necrotic neurons in rats with middle cerebral artery occlusion. Am J Pathol (in press)

Ginsberg MD, Stemau LL, Globus MYT, Dietrich WD, Busto R (1992) Therapeutic modulation of brain temperature: relevance to ischemic brain injury. Cerebrovasc Brain Metab Rev 4:189–225

Giulian D (1993) Reactive glia as rivals in regulating neuronal survival. Glia 7:102–110

Giulian D, Vaca K (1993) Inflammatory glia mediate delayed neuronal damage after ischemia in the CNS. Stroke 24:184–190

Graves AB, White E, Koepsell TD, Reifler BV, Van Bell G, Larson EB, Raskind M (1990) The association between head trauma and Alzheimer's disease. Am J Epidemiol 131:491–501

Griffin WS, Stanlet LC, Ling C, White L, MacLeod V, Perrot LJ, White CL, Araoz C (1989) Brain interleukin 1 and S-100 immunoreactivity are elevated in Down syndrome and Alzheimer's disease. Proc Natl Acad Sci USA 86:7611–7615

Griffin WS, Sheng JG, Roberts GW, Mrak RE (1995) Interleukin-1 expression in different plaque types in Alzheimer's disease: significance in plaque evolution. J Neuropathol Exp Neurol 54:276–281

Gross EC, Bednar MM, Howard DB, Sporn MB (1993) Transforming growth factor-β1 reduces infarct size after experimental cerebral ischemia in a rabbit model. Stroke 24:558–562

Gutierrez EG, Banks WA, Kastin AJ (1993) Murine tumor necrosis factor alpha is transported from blood to brain in the mouse. J Neuroimmunol 47:169–176

Gutierrez EG, Banks WA, Kastin AJ (1994) Blood-borne interleukin-1 receptor antagonist crosses the blood brain barrier. J Neuroimmunol 55:153–160

Henrich-Noack P, Prehn JHM, Krieglstein J (1994) Neuroprotective effects of TGF-β1. J Neural Transm 43:33–45

Hopkins SJ, Rothwell NJ (1995) Cytokines in the nervous system I: expression and receptors. TINS 18:83–88

Huberman M, Shalit F, Roth-Deri 1, Gutman B, Brodie C, Kott E, Sredni B (1994) Correlation of cytokine secretion by mononuclear cells of Alzheimer patients and their disease stage. Neuroimmunology 52:147–152

Joseph J, Grun JL, Lublin FD, Knobler RL (1993) Interleukin-6 induction in vitro in mouse brain endothelial cells and astrocytes by exposure to mouse hepatitis virus (MHV-4,JHM). J Neuroimmunol 42:47–52

Lawrence CB, Rothwell NJ (1994) Interleukin-1 receptor antagonist inhibits NMDA and AMPA receptor induced brain damage in the rat. Br J Pharmacol 112:484P

Lawrence CB, Strijbos PJLM, Rothwell NJ (1996) Role of interleukin-1 in excitotoxic neurodegeneration in vivo and in vitro. Eur J Neurosci (in press)

Lee SC, Liu W, Dickson DW, Brosren CF, Berman JW (1993) Cytokine production by human fetal microglia and astrocytes: differential induction by lipopolysaccharide and IL-1β. J Immunol 150:2659–2667

Liu T, McDonnell PC, Young PR, White RF, Siren AL, Hallenbeck JM, Barone FC, Fuerestein GZ (1993) Interleukin-1 beta MRNA expression in ischaemic rat cortex. Stroke 24:1746–1750

Loddick S, Rothwell NJ (1996) Neuroprotective effects of human recombinant interleukin-1 receptor antagonist in focal cerebral ischaemia in the rat. J Cerebr Blood Flow Metab (in press)

Logan A, Berry M, Gonzalez AM, Frautschy SA, Sporn MB, Baird A (1994) Effects of transforming growth factor beta 1 on scar production in the injured central nervous system of the rat. Eur J Neurosci 6:355–363

Luheshi GN, Gay J, Rothwell NJ (1994) Circulating IL-6 is transported into the brain via a saturable transport mechanism in the rat. Br J Pharmacol 111:146P

Lyons MK, Anderson RE, Meyer FB (1991) Corticotrophin releasing factor antagonist reduces ischaemic hippocampal neuronal injury. Brain Res 545:339–342

Margolis RL, Chuang DM, Post RM (1994) Programmed cell death: implications for neuropsychiatric disorders. Biol Psychiatry 35:946–956

Martin D, Chinookoswong N, Miller G (1994) The interleukin-1 receptor antagonist (rhIL-1ra) protects against cerebral infarction in a rat model of hypoxia-ischemia. Exp Neurol 130:362–267

Mattson MP (1990) Excitatory amino acids, growth factors and calcium: a teeter-totter model for neural plasticity and degeneration. Adv Exp Med Biol 268:211–220

Mattson MP, Cheng B (1993) Growth factors protect neurons against excitotoxic/ischaemic damage by stabilising calcium homeostasis. Stroke 24:1136–1140

Mayeux R, Ottman R, Tang MX, Noboa-Bauza L, Marder K, Gurland B, Stem Y, Gertrude H (1993) Genetic susceptibility and head injury as risk factors for Alzheimer's disease among community-dwelling elderly persons and their first degree relatives. Ann Neurol 33:494–501

McClain CJ, Cohen D, Ott L, Dinarello CA, Young B (1987) Ventricular fluid interleukin-1 activity in patients with head injury. J Lab Clin Med 15:48–54

McGeer PL, Akiyama H, Itagaki S, McGeer EG (1989) Immune system response in Alzheimer's disease. Can J Neurol Sci 16:516–527

McGeer PL, Itagaki S, Tago H (1987) Reactive microglia in patients with senile dementia of the Alzheimer's type are positive for the histocompatability glycoprotein HLA-DR. Neurosci Lett 79:195–200

Meda L, Cassatelia MA, Szendrei GI, Otvos L Jr, Baron P, Villalba M, Ferrari D, Rossi F (1995) Activation of microglial cells by β-amyloid protein and interferon. Nature 374:647–650

Minami M, Kuraishi Y, Yabuuchi K, Yamasaki A, Sato M (1992) Induction of interleukin-1 in rat brain after transient forebrain ischaemia. J Neurochem 58:390–392

Nicholson DW, Ali A, Thomberry NA, Vaillancourt JP, Ding CK, Gallant M, Gareau Y, Griffin PR, Labelle M, Lazenbnik YA, Munday NA, Raju SM, Smulson ME, Yamin TT, Yu VL, Miller DK (1995) Identification and inhibition of the ICE/CED-3 protease necessary for mammalian apoptosis. Nature 376:37–43

Ohyagi Y, Tabira T (1993) Effect of growth factors and cytokines on expression of amyloid beta protein precursor mRNAs in cultured neural cells. Mol Brain Res 18:127–132

Oppenheim RW (1991) Cell death during development of the nervous system. Annu Rev Neurosci 14:453–501

Panitch AS (1991) Interferons in multiple sclerosis. Drugs 44:946–962

Piani D, Frei K, Do KQ, Cuenod M, Fontanta A (1991) Murine brain macrophages induce NMDA receptor mediated neurotoxicity in vitro by secreting glutamate. Neurosci Lett 133:159–162

Pirtilla T, Hehta PD, Frey H, Wisniewski HM (1994) Alpha 1-antichymotrypsin and IL-1 beta are not increased in CSF or serum in Alzheimer's disease. Neurobiol Aging 15:313–317

Relton JK, Rothwell NJ (1992) Interleukin-1 receptor antagonist inhibits ischemic and excitotoxic neuronal damage in the rat. Brain Res Bull 29:243–246

Relton JK, Martin D, Russel D (1993) Effects of peripheral IL-lra treatment after focal ischaemia in the rat. Soc Neurosci Abs 19:673.2

Rivier C (1993) Neuroendocrine effects of cytokines in the rat. Rev Neurosci 4:223–237

Rothwell NJ, Hopkins SJ (1995) Interactions between cytokines and the nervous system II: actions and mechanisms. TINS 18:130–136

Rothwell NJ, Relton JK (1993) Involvement of interleukin-1 and lipocortin-1 in ischaemic brain damage. Brain Metab Rev 5:178–198

Royston MC, Rothwell NJ, Roberts GW (1992) Alzheimer's disease: from molecular biology to treatment. TIPS 13:131–133

Singh VK (1994) Studies of neuroimmune markers in Alzheimer's disease. Mol Neurobiol 9:73–81

Strijbos P, Rothwell NJ (1995) Interleukin-1 attenuates excitatory amino acid induced neurodegeneration in vitro: involvement of nerve growth factor. J Neurosci 13:179–185

Strijbos PJLM, Relton JK, Rothwell NJ (1994) Corticotrophin-releasing factor antagonist inhibits neuronal damage induced by focal cerebral ischaemia or activation of NMDA receptors in the rat brain. Brain Res 656:405–408

Toulmond S, Rothwell N (1995) Interleukin-1 receptor antagonist inhibits neuronal damage caused by fluid percussion injury in the rat. Brain Res 671: 261–266

Vandenabeele P, Fiers W (1991) Is amyloidogenesis during Alzheimer's disease due to an IL-1/IL-6 mediated acute phase response in the brain? Immunol Today 12:207–219

Vasilakos JP, Carroll RT, Emmerling MR, Doyle PD, Davis RE, Kim KS, Shivers BD (1994) Interleukin-1 beta dissociates beta-amyloid precursor protein and beta-amyloid peptide secretion. FEBS Lett 354:289–292

Wood JA, Wood PL, Ryan R, Graff-Radford NR, Pilapil C, Robitaille Y, Quirion R (1993) Cytokine indices in Alzheimer's temporal cortex: no changes in mature IL-1 beta or IL-lra but increases in the associated acute phase proteins IL-6, alpha 2-macroglobulin and C-reactive protein. Brain Res 629:245–252

Yabuuchi K, Minami M, Katsumata S, Yamazaki A, Satoh M (1994) An in situ hybridisation study on interleukin-1β mRNA induced by transient forebrain ischaemia in the rat brain. Mol Brain Res 26:135–142

Yamasaki Y, Matsuo Y Onodera H, Kogure K (1994) Interaction of neutrophils and endothelial cells in the pathogenesis of transient ischaemia. In: Krieglstein J, Oberpichler-Schwenk H (eds) Pharmacology of cerebral ischaemia. Wissenschaftliche Verlagsgesellschaft, Stuttgart, pp 427–435

Yamasaki Y, Matsuura N, Shozuhara H, Onodera H, Itoyama Y, Kogure K (1995) Interleukin-1 as a pathogenetic mediator of ischemic brain damage in rats. Stroke 26:676–681

Yang F, Jansen L, Friedrichs WE, Buchanan JM, Bowman BH (1993) IL-1 beta decreases expression of amyloid precursor protein gene in human glioma cells. Biochem Biophys Res Commun 191:1014–1019

Yuan J, Shaham S, Ledoux S, Ellis HM, Horvitz HR (1993) The *C. elegans* cell death gene ced-3 encodes a protein similar to mammalian interleukinβ converting enzyme. Cell 75:641–652

9 Inflammation in the CNS and in Alzheimer's Disease

V.H. Perry, M.D. Bell and D. Anthony

9.1 Introduction

The cellular response to neuronal injury or neuronal degeneration in the central nervous system (CNS) is commonly referred to as gliosis. Traditionally, this term was used to refer to the response by astroglia but in more recent years has come to refer to the whole spectrum of cellular reactions to neuronal damage. Since the glia, astroglia, oligodendroglia and microglia are the non-neuronal constituents of the CNS, the term gliosis suggests that this response is wholly intrinsic to the CNS compartment without a contribution from blood-derived leucocytes. In other tissues cellular degeneration is accompanied by an acute innate inflammatory response and leucocyte recruitment from the blood but in the CNS the term inflammation appears to have been widely used only in conditions where there is also an overt leucocyte infiltrate such as that seen in multiple sclerosis or viral infections of the CNS. Is it really the case that there is no inflammatory response when neurones degenerate? Is neuronal degeneration accompanied by a more subtle form of inflammation than that seen in other tissues?

It may seem irrelevant whether we refer to the non-neuronal response to CNS degeneration as gliosis or inflammation but we believe that this is an important distinction. A useful parallel exists with the recognition that the microglia are not simply another glial cell of neuroectodermal origin but are the resident mononuclear phagocytes in the CNS parenchyma (Streit et al. 1988; Perry and Gordon 1991; Perry 1994). The application of many of the ideas and tools of modern leucocyte and macrophage biology has resulted in a rapid increase in our knowledge about these cells and whether they are similar or different to other macrophage populations (Perry 1994; Gehrmann et al. 1995). Numerous laboratories are now studying their potential functions and their responses to injury in a wide variety of neuropathological conditions.

Investigations into whether there is or is not a typical inflammatory response in the CNS should serve to focus attention on the possible contribution of inflammation to the pathogenesis of chronic neurodegenerative diseases such as Alzheimer's disease (AD). In non-neuronal tissues it is well recognised that the acute inflammatory response is essential for both host defence and tissue repair (Movat 1985; Clark 1989). On the other hand, if the inflammatory response is not tightly controlled this may lead to non-specific tissue damage (Weiss 1989). We review here recent evidence showing that the inflammatory response in the CNS is quite unlike that in other tissues and in the light of this evidence review data implicating inflammation in the pathogenesis of AD.

9.2 Acute Inflammation in the CNS

Cellular degeneration in a non-neuronal tissue sets a complex sequence of events in motion resulting in a stereotyped recruitment of myelomonocytic cells from the blood. Neutrophils are recruited within minutes, reaching a peak rate at about 6 h with maximal numbers at 24 h. Monocytes soon follow the neutrophils and over the next few days the longer surviving monocytes and macrophages predominate (Movat 1985). If the foreign antigen or stimulus persists then this may result in the initiation of an acquired immune response and the activation and recruitment of T-lymphocytes.

The acute inflammatory response in the CNS parenchyma is highly unusual. Intracerebral injection of excitotoxic substances such as the glutamate agonists, ibotenic or kainic acid, results in the rapid degeneration of neurones. Despite the rapid degeneration of neurones there are few neutrophils, little oedema and perivascular cuffing is absent and this gives the impression that the response to neuronal degeneration does not involve an inflammatory component (Coffey et al. 1990; Andersson et al. 1991; Marty et al. 1991). Monocytes are, however, recruited from the blood but when they enter the CNS they rapidly develop processes and appear indistinguishable from activated microglia by both morphological criteria and their antigen expression. Whole body irradiation or blockade with antibodies against the complement type 3 receptor on macrophages reduces the numbers of activated microglia (Coffey et al. 1990; Andersson et al. 1991). It should be noted that the absence of neutrophil recruitment into the brain happens even in the presence of a damaged blood–brain barrier and monocytes apparently cross an intact blood–brain barrier and transform to develop processes just as they do in the developing brain (Coffey et al. 1990; Andersson et al. 1991).

The recruitment of monocytes to the region of degenerating cell bodies in a kainic acid lesion does not imply that all degenerating neuronal tissue provides the appropriate stimuli. Wallerian degeneration, which follows the destruction of the cell body, is accompanied by microglia activation and an increase in their number but this appears to be an intrinsic response without monocyte recruitment (see Lawson et al. 1994 for references). Similarly, during the retrograde reaction of motor neurone cell bodies following a peripheral nerve injury there is marked microglia activation and an increase in their number around the neurones. This response appears to involve only the resident cells (Streit and Graeber 1993). Our recent studies show that during Wallerian degeneration of the optic nerve, unlike Wallerian degeneration in a peripheral nerve, the vascular endothelium does not express the adhesion molecules necessary for monocyte recruitment (Castano et al. 1996a).

Thus, in many forms of acute neuronal degeneration the non-neuronal response is dominated by cells of the mononuclear phagocyte lineage with varying degrees of involvement of resident microglia and recruited monocytes. In histological sections the cells to have the appearance of activated microglia. The hallmarks of acute inflammation, the presence of large numbers of neutrophils, oedema and perivascular cuffing are absent. Is this a peculiarity of the inflammatory response to neuronal degeneration or a property intrinsic to the CNS microenvironment?

9.3 Exploring the Acute Inflammatory Response

A useful approach to examine the similarities and differences between the acute inflammatory response in the CNS parenchyma and other tissues is to inject a pro-inflammatory substance into both sites. We initially chose to do this with lipopolysaccharide (LPS) (Andersson et al. 1992a) since there is abundant literature on LPS-induced inflammation in other organs of the body. These studies demonstrated several important points.

It was immediately apparent that the CNS parenchyma was remarkably resistant to LPS challenge. Doses of LPS (20–200 ng), which produce a significant inflammatory response in skin, produced a florid response in the meninges and ventricles but not in the parenchyma. A dose of 200 ng of LPS injected into the parenchyma rapidly activated the resident microglia but very few neutrophils were recruited and monocytes only after a delay of several days. Very high doses of LPS (2 μg) produced the same response but in addition a late phase response at 5–7 days after injection, a phenomenon not seen in other tissues. In this late phase neutrophils, monocytes and lymphocytes were recruited to the parenchyma, especially the white matter.

Even with a severe inflammatory challenge the acute inflammatory response in the CNS parenchyma is dominated by the presence and activation of cells of the mononuclear phagocyte lineage, macrophages and microglia (Andersson et al. 1992a; Montero-Menei et al. 1994). Unlike the meninges or the choroid plexus and ventricles the CNS parenchyma appears to have evolved mechanisms to prevent the entry of neutrophils. We have suggested that these mechanisms evolved because the neutrophil is potentially a highly damaging cell in the CNS (Perry et al. 1993a). Neutrophils are well known to play a role in tissue damage and oedema (Weiss 1989). Oedema in the CNS parenchyma and within the confines of the skull is likely to be particularly damaging.

Possible reasons for the lack of neutrophil recruitment and delay in monocyte recruitment have been discussed elsewhere (Andersson et al. 1992a; Perry et al. 1993a). In brief it could be (1) CNS endothelium has delayed or deficient expression of adhesion molecules necessary for leucocyte adhesion and transendothelial migration, (2) the failure of intrinsic cells, astrocytes and microglia to generate the appropriate pro-inflammatory cytokines, (3) the failure of the intrinsic cells to generate the appropriate chemoattractants (chemokines) for specific leucocyte populations, (4) the presence of cytokine or chemokine inhibitors or factors that inhibit leucocyte migration. These factors alone or in combination might play a part.

We can now rule out the lack of adhesion molecule expression on CNS endothelium as an explanation for poor neutrophil recruitment. The selectins are rapidly upregulated following acute neuronal degeneration or LPS challenge (Bell and Perry, unpublished) and are also present on human CNS endothelium (Vora et al. 1995). The integrin ligands intercellular adhesion molecule (ICAM), vascular cell adhesion molecule (VCAM), and platelet endothelial cell adhesion molecule (PECAM), members of the immunoglobulin family, are essential for adherence and transmigration (Granger and Kubes 1994). ICAM and VCAM have also been demonstrated on CNS endothelium in both human CNS and in animal models (Cannella et al. 1990; Sobel et al. 1990; Engelhart et al. 1994). In the context of non-immune-mediated injury ICAM and VCAM are both readily expressed on CNS endothelium following an LPS challenge and following kainic acid-induced neuronal degeneration (Bell and Perry 1995). Their time course of expression was similar to that seen in peripheral tissues. We also found that PECAM was constitutively expressed on CNS endothelium at readily detectable levels but was not modulated by an inflammatory challenge.

To explore whether poor cytokine synthesis may account for the atypical CNS inflammatory response we injected a number of cytokines at high doses into the mouse or rat brain parenchyma (Andersson et al. 1992b; Anthony and Perry, in preparation). Although a prominent meningitis was produced following the injection of interleukin-1β (IL-1β) or tumour necrosis factor-α (TNF), there was little evidence of leucocyte recruitment to the CNS parenchyma. Thus, it is not the failure of LPS or excitotoxin-mediated neuronal degeneration to induce the cytokines that accounts for the lack of inflammation since the cytokines themselves are ineffective proinflammogens in the parenchyma. Intraparenchymal injections of chemokines, however, tell a different story. The chemokines are two related families of molecules that show remarkable specificity in their ability to chemoattract either neutrophils or mononuclear cells (Miller and Krangel 1992). The apparent intrinsic resistance of the CNS parenchyma to neutrophil recruitment can be overcome by the intraparenchymal injection (Bell et al. 1996) of recombinant macrophage inflammatory protein-2 (MIP-2; Wolpe et al. 1988) or interleukin-8 (Kunkel et al. 1991) with MIP-2 giving the greatest response. It is remarkable that a single chemokine, in the absence of other inflammatory mediators, can produce such a prominent and specific leucocyte influx into the CNS parenchyma. The mechanisms that regulate chemokine synthesis within the CNS, in particular those with neutrophil chemoattractant activity, are not known. This is particularly interesting in the light of data showing that LPS challenge to the neonatal brain not only results in a florid meningitis but also a prompt and marked influx of neutrophils and monocytes to the parenchyma (Lawson and Perry 1995). We do not know whether the differences between the inflammatory response in the neonatal brain and the adult brain can be accounted for by differential expression of chemokines. The postnatal regulation of inflammation in the brain parenchyma is an area with obvious clinical application.

The message that emerges from these studies on the leucocyte response to an acute inflammatory challenge highlight important differences between the brain parenchyma and other tissues. They also draw attention to distinct differences in the responses in different compartments of the CNS, the parenchyma contrasting with the choroid plexus/ventricles and the meninges.

9.4 Consequences of Acute Inflammation

There is a considerable amount of descriptive data documenting the presence of activated microglia in different models of CNS injury but their contribution to the outcome of these various pathologies is not well understood (Streit et al. 1988; Perry and Gordon 1991; Gehrmann et al. 1995). As the tools emerge that allow specific manipulation of different leucocyte populations it becomes possible to address these issues. In conditions where there is direct damage to the CNS vasculature neutrophils are recruited to the brain parenchyma. This happens in traumatic and ischemic injury (Clark et al. 1994; Chopp et al. 1994). Since neutrophils contribute to oedema and tissue damage in non-neuronal tissues (Weiss 1989) it is thus not entirely surprising that blockade of neutrophil recruitment in a model of cerebral ischaemia reduces the size of the lesion (Chopp et al. 1994).

The role of the macrophages and microglia is less well understood and also less easy to manipulate. A well-studied model of microglia activation is the facial nucleus paradigm (see Streit and Graeber 1993 for references). In this model transection of the facial nerve results in a cell body reaction in the neurones of the facial nucleus accompanied by microglia activation and proliferation. The microglia surround the cell bodies and appear to be involved in the removal of afferent synapses from the facial nucleus neurones and their processes; this may be a protective response to prevent the axotomized neurones from excess excitatory stimulation. Recent data cast some doubt on whether the microglia are essential for synaptic stripping, since prevention of microglia activation and proliferation by intraventricular infusion of cytosine arabinoside did not prevent synaptic stripping (Svensson and Aldskogius 1993a,b). It is interesting to note that microglia proliferation, in both the facial nucleus paradigm (Keifer and Kreutzberg 1991) and in Wallerian degeneration of the murine optic nerve (Castano et al. 1996b), was not significantly influenced by peripheral administration of dexamethasone, a treatment that is known to reduce proliferation of macrophages both in vivo and in vitro.

Giulian and colleagues have examined the role of macrophages and microglia in a model of spinal cord ischaemia (Giulian and Robertson 1990) and following a penetrating wound (Giulian et al. 1989). Following peripheral administration of a cocktail of colchicine and chloroquine there was behavioural sparing in the spinal ischaemia model (Giulian and Robertson 1990) and evidence of reduced phagocytosis following cerebral trauma (Giulian et al. 1989). The mechanism by which colchicine and chloroquine interfere with macrophage function to produce sparing of behavioural function is not known. It is interesting to note that in both of these models dexamethasone was also relatively ineffective at modulating the macrophages as described above in Wallerian degeneration and the retrograde response. The lack of influence of glucocorticoids on CNS macrophages and microglia is of interest in the context of a therapeutic treatment of CNS inflammation.

Another approach to investigating the functions of the mononuclear phagocytes is to target their secretory products with neutralizing antibodies or inhibitors. For example, it has been shown that IL-1β, a secretory product of activated macrophages, can exacerbate both ischaemia- and excitotoxin-induced neuronal injury, and interleukin receptor antagonist protein (IRAP) will reduce the lesion size (Relton and Rothwell 1992). Precisely how the IL-1β or IRAP produces these effects is at present unclear. It has been shown that following a penetrating wound to the brain the expression of TGF-β1 is induced in a number of different cell types around the wound including astrocytes and mononuclear phagocytes (Logan et al. 1992, 1994). Application of neutralising antibodies to the site of the wound prevents the formation of a fibrotic scar and application of TGF-β1 enhances scar formation (Logan et al. 1994). The contribution of TGF-β1 to scar formation is clear but the relevance of scar formation in CNS repair in general is unresolved.

One obvious capability of the mononuclear phagocytes is to remove the debris resulting from tissue damage. However, macrophages and microglia in the CNS do this rather ineffectually when compared to their peripheral counterparts. During Wallerian degeneration the axonal and myelin debris is cleared very rapidly from the peripheral nervous system (PNS) when compared to the CNS (Bignami et al. 1981) and in man the debris of Wallerian degeneration in the CNS may persist for years (Miklossy and Van der Loos 1991). This is despite the fact that the relative increase in PNS and CNS mononuclear phagocytes is comparable during Wallerian degeneration (Lawson and Perry 1995).

The routes to manipulate the macrophage/microglia response during an acute inflammatory response to neuronal degeneration in the CNS are at present largely uncharted territory. It is clear that if we are to understand the role of these cells new methodologies are required.

9.5 Inflammation in AD

It is against the background of the unusual acute inflammatory response in the CNS that we consider the evidence for the possible involvement of inflammation in the pathogenesis of AD. The hallmarks of the pathology of AD are the presence of plaques and tangles within the grey matter of the brain. Deposits of amyloid form immature or amorphous plaques which under certain conditions will progress to form mature plaques associated with dystrophic neurites. The neurofibrillary tangles within neurones are formed by filaments of hyperphosphorylated tau. The disease is progressive and the pathology spreads across the cortical mantle in a predictable fashion (Pearson et al. 1985; Braak et al. 1993). The possibility that an acute or chronic inflammatory response contributes to the onset or progression of the pathology is important since this offers a possible therapeutic approach for a disease in which there is at present no treatment.

9.6 Cellular Responses

Numerous reports have now documented the presence of activated microglia in AD brains. The activation of microglia is a part of the normal process of ageing in the CNS (Perry et al. 1993b; Rogers et al. 1988) but the changes observed in AD are greater than those in the aged brain. The microglia express enhanced levels of major histocompatibility complex (MHC) antigens, leucocyte common antigen (CD45), the β2-integrins, type 2 plasminogen activator inhibitor and a number of other molecules (Akiyama et al. 1993, 1994a; McGeer et al. 1987, 1993; Rozemuller et al. 1992; Tooyama et al. 1990). In general, the upregulation of these molecules is also present in other neurodegenerative pathologies and not specific to AD. The activated microglia are commonly found in association with mature rather than amorphous or immature plaques (Rozemuller et al. 1992). Both light microscope and electron microscope observations have given rise to the notion that the microglia may not just be involved in attempting to phagocytose the amyloid deposits but may actually be involved in the deposition of amyloid (Wegiel and Wisniewski 1990). This notion receives some support from data showing that leucocytes, including microglia, may synthesise a splice-variant of the amyloid precursor protein (Konig et al. 1992).

The expression of mononuclear phagocyte-associated secretory products by the activated microglia has not been well studied. There are reports that microglia in AD brains stain for the presence of TNF-α, IL-Iβ and IL-6 (Griffin et al. 1994; Bauer et al. 1992; Dickson et al. 1993). However, given the importance of these cytokines in the inflammatory process, it is somewhat surprising that their expression has not been more widely documented.

Apart from the microglia the only other leucocytes with an increased presence in the AD brain are the T-lymphocytes. There are two reports (Rogers et al. 1988; Itagaki et al. 1988) that both CD8 positive (CD8+) and CD4+ lymphocytes are increased in number in AD brains associated with blood vessels and also within the parenchyma. The CD8+ cells are reported to outnumber the CD4+ cells. The function of these cells in the AD brain would indicate the presence of an inflammatory response with an immune component rather than an innate inflammatory response. The presence of a T-lymphocyte population in the AD brain also requires further investigation.

9.7 Acute Phase Proteins

The acute phase response is a component of the body's defence against injury and infection. It occurs rapidly and before any antibody-mediated immune response. The response is initiated by cytokines and involves a complex series of physiological and metabolic changes including the synthesis of "acute phase proteins" or "acute phase reactants" in the liver (Steel and Whitehead 1993). The proteins that are synthesised include proteins of the complement cascade, fibrinogen, serum amyloid A protein (SAA), C-reactive protein (CRP) or serum amyloid P protein (SAP), a number of proteinase inhibitors including α1-antichymotrypsin (ACT). While these acute phase reactants have diverse and important functions in host defence some may also have damaging effects in chronic inflammatory disease (Steel and Whitehead 1993). The presence of acute phase proteins in the AD brain not only provides evidence for an inflammatory response but these proteins may also contribute to the pathogenesis.

Since the first report that several proteins of the complement cascade were associated with AD pathology (Eikelenboom and Stam 1982), numerous reports have confirmed and extended these observations (Eikelenboom et al. 1989; McGeer et al. 1989, 1993; McGeer and McGeer 1992). Complement activation is important because the complement cascade not only generates opsonins to enhance phagocytosis but also chemoattractants for leucocytes and the cell lytic membrane attack complex. The membrane attack complex, C5b-C9, in areas of dystrophic neurites and neurofibrillary tangles (McGeer et al. 1989) raises the possibility that complement-mediated neuronal killing may occur in AD. The proteins involved in protecting cells against complement-mediated lysis are also associated with dystrophic neurites (McGeer et al. 1993).

The stimulus for the complement activation is not obvious but the evidence suggests that it is activation of the classical complement pathway that is involved. Activation of the classical pathway is usually associated with immune complexes but there are many other possible activating molecular complexes including CRP bound to bacterial polysaccharides and mannan-binding proteins (Reid and Thiel 1993). There is recent evidence in vitro to show that the classical pathway may be activated by the amyloid β protein (Rogers et al. 1992).

The complement proteins like other acute phase proteins are largely synthesised by the liver but there is now good evidence to show extrahepatic sites of complement synthesis in particular by tissue macrophages (Whaley and Lemercier 1993). Thus, the presence of complement proteins in the CNS does not necessarily mean that the blood–brain barrier is damaged in AD (see below), and in the CNS the obvious candidate cells for the synthesis of components of the classical pathway are the resident mononuclear phagocytes, the microglia. Astrocytes may also synthesise some components of the complement pathway (see Eddleston and Mucke 1993). It has recently been shown that a human astrocyte-tumour cell line can synthesise and secrete all of the components of the activation pathway of complement and also produce components of the terminal lytic pathway (Gasque et al. 1995). The same authors showed that a human monocyte cell line can also produce all of the terminal components.

The degree to which complement activation and possible bystander lysis of neurones is particular to or a causal event in AD is unclear. Some evidence for complement activation being specific to AD comes from observations that the amyloid deposits in scrapie-inoculated mice did not have complement proteins associated with them (Eikelenboom et al. 1991). However, this deserves further investigation since recent evidence shows prominent microglia activation in scrapie (Williams et al. 1994). Complement components have also been shown to be rapidly expressed in the CNS parenchyma after a kainic acid lesion (Akiyama et al. 1994b) or during the retrograde cell body response to a peripheral nerve injury (Svensson and Aldskogius 1992). A problem in immunohistochemical studies such as those performed on AD material is that it is impossible to determine when a complex such as C5b-C9 was deposited. The presence of the C5b-C9 complex on dystrophic neurites could be related to the death of the neurites or simply be non-specific binding of a very large protein complex to an inert surface. Many details of the relationship between the complement proteins and AD pathology are not apparent from the literature.

Immunohistochemical techniques have identified other acute phase proteins in association with AD pathology, in particular in association with plaques. ACT has been localised to amyloid plaques and both protein and mRNA for ACT have been demonstrated in reactive astrocytes (Abraham et al. 1988; Abraham 1992; Pasternak et al. 1989). ACT is a serine protease inhibitor but the natural protease inhibited in the CNS has not yet been characterised. SAP (Kalaria 1992; Licastro et al. 1995; Perlmutter et al. 1995) and to a lesser extent CRP (Kalaria and Hedera 1995) have been shown to be associated with plaques. As with complement the significance of these protein deposits within the CNS and bound to plaques is difficult to interpret.

The presence of SAP bound to plaques is of particular interest since it has been shown to be synthesised in the liver but not in the CNS: SAP mRNA has been undetectable by polymerase chain reaction (PCR; Kalaria 1992). This observation suggests that although there is no generalised breakdown of the blood–brain barrier in AD, blood–brain barrier function may not be entirely normal. Support for this comes also from the fact that there is reduced expression of the blood–brain barrier glucose transporter (Kalaria and Harik 1989), and also capillaries with collapsed or degenerating endothelium have been described (Kalaria and Hedera 1995). The presence of a partially or transiently compromised blood–brain barrier has implications for how we interpret the presence of acute phase proteins in association with amyloid plaques.

9.8 Inflammation in AD:
Causal, Contributor or Consequence?

The role of the inflammatory process in the pathology of AD lies somewhere along the spectrum of causal – contributory – consequence. On the basis of the available information, is it possible to be more precise?

At the present time, there is little reason to believe that inflammation in the CNS provoked by some unknown agent or agents will lead to AD. There is some evidence that head injury, which is associated with activation of microglia and the synthesis of IL-1, is a risk factor in AD (Griffin et al. 1994) but this is likely to involve only a minority of cases of sporadic AD. It unclear how a non-specific head injury could predispose the generation of a pathology that is commonly initiated in the region of the hippocampus (Braak et al. 1993).

Evidence for a contributory role of inflammation comes from several tantalising studies which have examined the prevalence of AD in groups of patients taking anti-inflammatory drugs. A retrospective study based on autopsy records reported that AD was low in patients with rheumatoid arthritis, leading to the hypothesis that anti-inflammatory drugs have some influence on the onset of AD (McGeer et al. 1992a). This idea has recently received support from a twin study in which there was evidence that anti-inflammatory drugs delayed the onset of AD (Breitner et al. 1994). A further study showed that the prevalence of dementia in patients taking dapsone was lower than in an untreated group (McGeer et al. 1992b). Dapsone is a drug used to treat leprosy but also has anti-inflammatory properties. It is important to note that these effects on prevalence are seen after patients had taken the drugs for many years. It is, thus, unclear at precisely what stage of the disease they are likely to be acting.

A further piece of evidence indicating a contribution of inflammation to AD pathology comes from a small 6-month trial with indomethacin in patients suffering from AD (Rogers et al. 1993). The results showed delayed cognitive decline in the treated group when compared to the placebo group but only a small number of patients completed the trial due to the side effects of indomethacin.

9.9 Overview and Opportunities

The absence of a typical inflammatory response in the AD brain is entirely consistent with what we know of the atypical acute inflammatory response in the CNS. The presence of activated microglia, components of the complement cascade and other acute phase proteins all support the idea that there is an ongoing inflammatory response. These indicators do not tell us whether this is a response to the pathology or a contributory component. However, prevalence studies comparing populations of patients treated with anti-inflammatory drugs indicate that these drugs offer some protection against the onset and/or progression of the disease, indicating a contributory rather than consequential role.

The presence of an inflammatory response in the CNS which may contribute to the pathogenesis of AD is not only important from the point of view of a therapeutic target but also as a possible source of surrogate markers of disease progression or arrest during therapy. The measurement of cognitive decline or its arrest in a slowly progressive disease is a complex problem. A therapy that arrests the degenerative process might be expected to also have an effect on the inflammatory response. Detection of inflammation in the brain in life, by imaging (Ramsey et al. 1992), or by the analysis of body fluids is a possibility. However, it should be noted that the latter is unlikely to be straightforward since even in a CNS disease such as multiple sclerosis with a florid inflammatory component measurement of cytokines has not turned out to be a simple predictor of the diseased state (Maimone et al. 1991; Weller et al. 1991). The atypical nature of CNS inflammation suggests that measurement of typical inflammatory mediators is unlikely to be useful. The tight regulation of the cytokine network within the tissue is also unlikely to lead to the appearance of cytokines within the cerebrospinal fluid or blood. The contradictory results already in the literature (Fillit et al. 1991; Cacabelos et al. 1994) testify to the potential difficulties.

Much of the information available on inflammation or markers of inflammation in AD brains is fragmentary. The data in the literature give little idea of regional variations within the CNS, how the inflammatory response or its indicators evolve, or individual patient variation. The fact that the spread of the disease has a pattern (Pearson et al. 1985; Braak et al. 1993) and the demonstration that there is a rapid and catastrophic loss of tissue from the medial temporal lobe of AD patients (Jobst et al. 1994) suggest that it would be important to evaluate markers of inflammation with these reference points in mind as well as the time course of the individual's disease.

There is a body of evidence to show that microglia are activated but it is much less clear what the consequences of this activation might be. The macrophage has the potential to secrete an extraordinary range of inflammatory mediators, inflammatory modulators and growth factors (Rappolee and Werb 1992), few of which have been explored in the context of AD pathology. However, as is clear from results discussed above, the effects of particular cytokines within the CNS microenvironment may not be the same as those seen in peripheral tissues: simple extrapolation from what we know of their peripheral effects to the CNS is probably too simplistic. We know little about the mechanisms that are involved in the activation of microglia although there is evidence to implicate macrophage colony stimulating factor-1 (Raivich et al. 1994), the complement type 3 receptor (Reid et al. 1993) and the NF-κB transcription factor (Kaltschmidt et al. 1994). Even less is known about what keeps microglia normally quiescent. If CNS inflammation is to be a therapeutic target these are clearly important issues to be addressed.

A major problem for the study of AD pathology and the possible importance of an inflammatory component is the absence of animal models in which some of the ideas could be tested. The majority of animal models of CNS disease are designed to address problems of an acute nature rather than a chronic, slowly evolving disease. It has been pointed out that the spongiform encephalopathies of animals and man have many similarites with AD and that insights arising from study of these diseases may provide important insights for AD research (DeArmond 1993). The generation of transgenic mice which develop some aspects of AD pathology is also likely to be a valuable approach (Games et al. 1995).There is clearly much to do but the potential benefits are obvious.

Acknowledgments. The work in the authors' laboratory was supported by the Wellcome Trust and the Multiple Sclerosis Society.

References

Abraham CR (1992) The role of the acute-phase protein a-1 antichymotrypsin in brain dysfunction and injury. Res Immunol 143:631–636

Abraham CR, Selkoe DJ, Potter H (1988) Immunohistochemical identification of the serine protease inhibitor alpha-1 antichymotrypsin in the brain amyloid deposits of Alzheimer's disease. Cell 52:487–501

Akiyama H, Ikeda K, Kondo H, Kato M, McGeer PL (1993) Microglia express the type 2 plasminogen activator inhibitor in the brain of control subjects and patients with Alzheimer's disease. Neurosci Lett 164:233–235

Akiyama H, Ikeda K, Katoh M, McGeer EG, McGeer PL (1994a) Expression of MRP14, 27E10, interferon-alpha and leukocyte common antigen by reactive microglia in postmortem human brain tissue. J Neuroimmunol 50:195–201

Akiyama H, Tooyama I, Kondo H, Ikeda K, Kimura H, McGeer EG, McGeer PL (1994b) Early response of brain resident microglia to kainic acid-induced hippocampal lesions. Brain Res 635:257–268

Andersson P-B, Perry VH, Gordon S (1991) The kinetics and morphological characteristics of the macrophage-microglial response to kainic acid-induced neuronal degeneration. J Neurosci 42:201–214

Andersson, P-B, Perry, VH, Gordon S (1992a) The acute inflammatory response to lipopolysaccharide in CNS parenchyma differs from that in other body tissues. Neuroscience 48:169–186

Andersson P-B, Perry VH, Gordon S (1992b) Intracerebral injection of proinflammatory cytokines or leukocyte chemotaxins induces minimal myelomonocytic cell recruitment to the parenchyma of the central nervous system. J Exp Med 176:255–259

Bauer J, Ganter U, Strauss S, Stadtmuller G, Frommberger U, Bauer H, Volk B, Berger M (1992) The participation of interleukin-6 in the pathogenesis of Alzheimer's disease. Res Immunol 143:650–657

Bell MD, Perry VH (1995) Adhesion molecule expression on murine cerebral endothelium following the injection of a proinflmmogen or during acute neuronal degeneration. J Neurocytol 24:695–710

Bell MD, Taub DD, Perry VH (1996) Overriding the brain's intrinsic resistance to leukocyte recruitment with intraparenchymal injections of recombinant chemokines. Neuroscience (in press)

Bignami A, Dahl D, Nguyen BT, Crosby CJ (1981) The fate of axonal debris in Wallerian degeneration of rat optic and sciatic nerves. J Neuropathol Exp Neurol 40: 337–350

Braak H, Braak E, Bohl J (1993) Staging of Alzheimer-related cortical destruction. Eur Neurol 33:403–408

Breitner JCS, Gau BA, Welsh KA, Plassman BL, McDonald WM, Helms MJ, Anthony JC (1994) Inverse association of anti-inflammatory treatments and Alzheimer's disease: initial results of a co-twin control study. Neurology 44:227–232

Cacabelos R, Alvarez XA, Franco-Maside A, Fernandez-Novoa L, Caamano J (1994) Serum tumour necrosis factor (TNF) in Alzheimer's disease and multi-infarct dementia. Methods Find Exp Clin Pharmacol 16:29–35

Cannella B, Cross AH, Raine CS (1990) Upregulation and co-expression of adhesion molecules correlate with relapsing autoimmune demyelination in the the central nervuos system. J Exp Med 172: 1521–1524

Castano A, Bell MD, Perry VH (1996a) Unusual aspects of inflammation in the nervous system: Wallerian degeneration. Neurobiol Aging (in press)

Castano A, Lawson LJ, Fearn S, Perry VH (1996b) Activation and proliferation of murine microglia are insensitive to glucocorticoids in Wallerian degeneration. Eur J Neurosci (in press)

Chopp WM, Zhang RL, Chen H, Li Y, Jiang N, Rusche JR (1994) Postischemic administration of an anti-Mac-1 antibody reduces ischemic cell damage after transient middle cerebral artery occlusion in rats. Stroke 25: 869–75

Clark RAF (1989) Wound repair. Curr Opin Cell Biol 1:1000–1008

Clark RSB, Schiding JK, Kaczorowski SL, Marion DW, Kochanek PM (1994) Neutrophil accumulation after traumatic brain injury in rats: comparison of weight drop and controlled cortical impact models. J Neurotrauma 5:499–506

Coffey PJ, Perry VH, Rawlins JNP (1990) An investigation into the early stages of the inflammatory response following ibotenic acid-induced neuronal degeneration. Neuroscience 35:121–132

DeArmond SJ (1993) Alzheimer's disease and Creutzfeldt-Jakob disease: overlap of pathogenic mechanisms. Curr Opin Neurol 6:872–881

Dickson DW, Lee SC, Mattiace LA, Yn SH, Brosnan C (1993) Microglia and cytokines in neurological disease, with special reference to AIDS and Alzheimer's disease. Glia 7:75–83

Eddleston M, Mucke L (1993) Molecular profile of reactive astrocytes – implications for their role in neurologic disease. Neuroscience 54:15–36

Eikelenbloom P, Stam FC (1982) Immunoglobulins and complement factors in senile plaques: an immunohistochemical study. Acta Neuropathol (Berl) 245:417–420

Eikelenbloom P, Hack CE, Rozemuller JM, Stam FC (1989) Complement activation in amyloid plaques in Alzheimer's dementia. Virchows Arch [B] Cell Pathol 56:259–262

Eikelenboom P, Rozemuller JM, Kraal G, Stam FC, McBride PA, Bruce ME, Fraser H (1991) Cerebral amyloid plaques in Alzheimer's disease but not scrapie-affected mice are closely associated with a local inflammatory process. Virchows Arch [B] 60:329–331

Engelhardt B, Conley FK, Butcher EC (1994) Cell-adhesion molecules on vessels during inflammation in the mouse central nervous system. J Neuroimmunol 51:199–208

Fillit H, Ding WH, Buee L, Kalman J, Altsteil L, Lawlor B, Wolf-Klein G (1991) Elevated circulating tumour necrosis factor levels in Alzheimer's disease. Neurosci Lett 129:318–320

Games D, Adams D, Alessandrini R et al (1995) Alzheimer's type neuropathology in transgenic mice overexpressing V717F β-amyloid precursor protein. Nature 373:523–527

Gasque P, Fontaine M, Morgan BP (1995) Complement expression in human brain. Biosynthesis of terminal pathway components and regulators in human glial cells and cell lines. J Immunol 154:4726–4733

Gehrmann J, Matsumoto Y, Kreutzberg GW (1995) Microglia: intrinsic immunoeffector cell of the brain. Brain Res Rev 20:269–287

Giulian D, Robertson C (1990) Inhibition of mononuclear phagocytes reduces ischemic injury in the spinal cord. Ann Neurol 27:33–42

Giulian D, Chen J, Ingeman JE, George JK, Naponen M (1989) The role of mononuclear phagocytes in wound healing after traumatic injury to adult mammalian brain. J Neurosci 9:4416–4429

Granger DN, Kubes P (1994) The microcirculation and inflammation: modulation of leucocyte-endothelial cell adhesion. J Leukoc Biol 55:662–675

Griffin WS, Sheng JG, Gentleman SM, Graham DI, Mrak RE, Roberts GW (1994) Microglial interleukin-1 alpha expression in human head injury: correlations with neuronal and neuritic beta-amyloid precursor protein expression. Neurosci Lett 176:133–136

Itagaki S, McGeer PL, Akiyama H (1988) Presence of T-cytotoxic suppressor and leukocyte common antigen positive cells in Alzheimer disease brain tissue. Neurosci Lett 91:259–264

Jobst KA, Smith AD, Szatmari M, Esiri MM, Jaskowski A, Hindley N, McDonald B, Molyneux AJ (1994) Rapidly progressing atrophy of medial temporal lobe in Alzheimer's disease. Lancet 343:829–830

Kalaria RN (1992) Serum amyloid P and related molecules associated with the acute-phase response in Alzheimer's disease. Res Immunol 143:637–645

Kalaria RN, Harik SI (1989) Reduced glucose transporter at the blood–brain barrier in cerebral cortex in Alzheimer disease. J Neurochem 53:1083–1088

Kalaria RN, Hedera P (1995) Differential degeneration of the cerebral microvasculature in Alzheimer's disease. Neuroreport 6:477–480

Kaltschmidt C, Kaltschmidt B, Lannes-Viera J, Kreutzberg GW, Wekerle H, Bauerle PA, Gehrmann J (1994) Transcription factor NF-kB is activated during experimental allergic encephalomyelitis. J Neuroimmunol 55:99–106

Keifer R, Kreutzberg GW (1991) Effects of dexamethasone on microglia activation in vivo: selective downregulation of major histocompatibility complex class II expression in regenerating facial nucleus. J Neuroimmunol 34:99–108

Konig G, Monning U, Czech C, Prior R, Banati R, Schreiter-Gasser U, Bauer J, Masters CL, Beyreuther K (1992) Identification and differential expression of a novel alternative splice isoform of the beta A4 amyloid precursor protein (APP) mRNA in leucocytes and brain microglial cells. J Biol Chem 267:10804–10809

Kunkel S, Standiford T, Kasahara K, Streiter R (1991) Interleukin-8 (IL-8): the major neutrophil chemotactic factor in the lung. Exp Lung Res 17:17–23

Lawson LJ, Perry VH (1995) The unique characterisitics of inflammatory responses in mouse brain are acquired during postnatal development. Eur J Neurosci 7:1584–1595

Lawson LJ, Frost L, Risbridger J, Fearn S, Perry VH (1994) Quantification of the mononuclear phagocyte response to Wallerian degeneration of the optic nerve. J Neurocytol 23:729–744

Licastro F, Morini MC, Polazzi E, Davis LJ (1995) Increase serum alpha 1-antichymotrypsin in patients with probable Alzheimer's disease: an acute phase reactant without the peripheral acute phase response. J Neuroimmunol 57:71–75

Logan A, Frautschy SA, Gonzalez A-M, Sporn MB, Baird A (1992) Enhanced expression of transforming growth factor 1 in the rat brain after a localized cerebral injury. Brain Res 587:216–225

Logan A, Berry M, Gonzalez A-M, Frautschy SA, Sport MB, Baird A (1994) Effects of transforming growth factor 1 on scar production in the injured central nervous system of the rat. Eur J Neurosci 6:355–363

Maimone D, Gregory S, Amason BG, Reder AT (1991) Cytokine levels in the cerebrospinal fluid and serum of patients with multiple sclerosis. J Neuroimmunol 32:67–74

Marty S, Dusart I, Peschanski M (1991) Glial changes following an excitotoxic lesion in the CNS-I microglia/macrophages. Neuroscience 45:529–539

McGeer PL, McGeer EG (1992) Complement proteins and complement inhibitors in Alzheimer's disease. Res Immunol 143:621–624

McGeer PL, Itagaki S, Tago H, McGeer EG (1987) Reactive microglia in patients with senile dementia of the Alzheimer type are positive for the histocompatibility glycoprotein HLA-DR. Neurosci Lett 79:195–200

McGeer PL, Akiyama H, Itagaki S, McGeer EG (1989) Activation of the classical complement pathway in brain tissue of Alzheimer patients. Neurosci Lett 107:341–346

McGeer PL, McGeer EG, Rogers J, Sibley J (1992a) Does anti-inflammatory treatment protect against Alzheimer's disease? In: Khachaturian ZS, Blass JP (eds) Alzheimer's disease: new treatment strategies. Dekker, New York, pp 165–171

McGeer PL, Harada N, Kimura H, McGeer EG, Schulzer M (1992b) Prevalence of dementia amongst elderly Japanese with leprosy: apparent effect of chronic drug therapy. Dementia 3:146–149

McGeer PL, Kawamata T, Walker DG, Akiyama H, Tooyama I, McGeer E (1993) Microglia in degenerative neurological disease. Glia 7:84–92

Miklossy J, van der Loos H (1991) The long distance effects of brain lesions: visualization of myelinated pathways in the human brain using polarizing and fluorescence microscopy. J Neuropathol Exp Neurol 50:1–15

Miller M, Krangel M (1992) Biology and biochemistry of the chemokines: a family of chemotactic and inflammatory cytokines. Crit Rev Immunol 12:17–46

Montero-Menei CN, Sindji L, Pouplard-Barthelaix A, Jehan F, Denechaud L, Darcy F (1994) Lipopolysaccharide intracerebral administration induces minimal inflammatory reaction in rat brain. Brain Res 653: 101–11

Movat HZ (1985) The inflammatory reaction. Elsevier, Amsterdam

Pasternak JM, Abraham CR, Van Dyke B, Potter H, Younkin SG (1989) Astrocytes in Alzheimer's disease gray matter express a1-antichymotrypsin mRNA. Am J Pathol 135:827–834

Pearson RC, Esiri MM, Hiorns RW, Wilcock GK, Powell TP (1985) Anatomical correlates of the distribution of the pathological changes in the neocortex in Alzheimer disease. Proc Natl Acad Sci USA 82:4531–4534

Perlmutter LS, Barron E, Myers M, Saperia D, Chui HC (1995) Localization of amyloid P component in human brain: vascular staining patterns and association with Alzheimer's disease lesions. J Comp Neurol 352:92–105

Perry VH (1994) Macrophages and the nervous system. Molecular biology intelligence unit. Landes, Austin, Texas

Perry VH, Gordon S (1991) Macrophages and the nervous system. Int Rev Cytol 125:203–244

Perry VH, Andersson PB, Gordon S (1993a) Macrophages and inflammation in the central nervous system. Trends Neurosci 16:268–273

Perry VH, Matyszak MK, Fearn S (1993b) Altered antigen expression of microglia in the aged rodent CNS. Glia 7:60–67

Raivich G, Moreno-Flores MT, Moller JC, Kreutzberg GW (1994) Inhibition of posttraumatic microglial proliferation in a genetic model of macrophage colony-stimulating factor deficiency in the mouse. Eur J Neurosci 6:1615–1618

Ramsey SC, Weiller C, Myers R, Cremer JE, Luthra SK, Lammertsma AA, Frachowiak RS (1992) Monitoring by PET of macrophage accumulation in brain after ischemic stroke. Lancet 339:1054–1055

Rappolee DA, Werb Z (1992) Macrophage-derived growth factors. In: Russel WS, Gordon S (eds) Macrophage biology and activation. Springer, Berlin Heidelberg New York, pp 87–140 (Current topics in microbiology and immunology, vol 181)

Reid KMB, Thiel S (1993) C1q and related molecules in defence. In: Sim E (ed) The natural immune system: humoral factors. IRL Press, Oxford, pp 151–175

Reid DM, Perry VH, Andersson P-B, Gordon S (1993) Mitosis and apoptosis of microglia in vivo induced by an anti-CR3 antibody which crosses the blood–brain barrier. Neuroscience 56:529–533

Relton JK, Rothwell NJ (1992) Interleukin-1 receptor antagonist inhibits ischaemic and excitotoxic neuronal damage in the rat. Brain Res Bull 29:243–246

Rogers J, Luber-Narod J, Styren SD, Civin WH (1988) Expression of immune system-associated antigen by cells of the human central nervous system: relationship to the pathology of Alzheimer disease. Neurobiol Aging 9:339–349

Rogers J, Cooper NR, Webster S, Schultz J, McGeer PL, Styren SD, Civin WH, Brachova L, Bradt B, Ward P, Lieberberg I (1992) Complement activation by β-amyloid in Alzheimer disease. Proc Natl Acad Sci USA 89:10016–10020

Rogers J, Kirby LC, Hempelman SR, Berry DL, McGeer PL, Kaszniak AW, Zalinski J, Cofield M, Mansukhani L, Willson P, Kogan F (1993) Clincial trial of indomethacin in Alzheimer's disease. Neurology 43:1609–1611

Rozemuller JM, van der Valk P, Eikelenboom P (1992) Activated microglia and cerebral amyloid deposits in Alzheimer's disease. Res Immunol 143:646649.

Sobel RA, Mitchell ME, Fonfren G (1990) Intercellular adhesion molecule-1 (ICAM-1) in cellular immune reactions in the human central nervous system. Am J Pathol 136:1309–1316

Steel DM, Whitehead AS (1993) The acute phase response. In: Sim E (ed) The natural immune system: humoral factors. IRL Press, Oxford, pp 1–29

Streit WJ, Graeber MB (1993) Heterogeneity of microglia and perivascular cell populations: insights gained from the facial nucleus paradigm. Glia 7:68–74

Streit WJ, Graeber MB, Kreutzberg GW (1988) Functional plasticity of microglia: a review. Glia 1:301–307

Svensson M, Aldskogius H (1992) Evidence for activation of the complement cascade in the hypoglossal nucleus following peripheral nerve injury. J Neuroimmunol 40:99–109

Svensson M, Aldskogius H (1993a) Infusion of cytosine-arabinoside into the cerebrospinal fluid of the rat brain inhibits the microglial cell proliferation after hypoglossal nerve injury. Glia 7:286–298

Svensson M, Aldskogius H (1993b) Synaptic density of axotomized hypoglossal motorneurons following pharmacological blockade of the microglial cell proliferation. Exp Neurol 120:123–131

Tooyama I, Kimura H, Akiyama H, McGeer PL (1990) Reactive microglia express class I and class II major histocompatibility complex antigens in Alzheimer's disease. Brain Res 523:273–280

Vora AJ, Perry ME, Hobbs C, Dumonde DC, Brown KA (1995) Selective binding of peripheral blood lymphocytes to the walls of cerebral blood vessels in frozen sections of human brain. J Immunol Methods 180:165–180

Wegiel J, Wisniewski HM (1990) The complex of microglial cells and amyloid star in three-dimensional reconstruction. Acta Neuropathol (Berl) 81:116–124

Weiss SJ (1989) Tissue destruction by neutrophils. N Engl J Med 320:365–376

Weller M, Stevens A, Sommer N, Melms A, Dichgans J, Wietholter H (1991) Comparative analyis of cytokine patterns in immunological, infectious, and oncological neurological disorders. J Neurol Sci 104:215–221

Whaley K, Lemercier C (1993) The complement system. In: Sim E (ed) The natural immune system: humoral factors. IRL Press, Oxford, pp 120–150

Williams AE, Lawson LJ, Perry VH, Fraser H (1994) Characterization of the microglial response in murine scrapie. Neuropathol Appl Neurobiol 20:47–55

Wolpe S, Davatelis G, Sherry G, Beutler B, Hesse D, Hguyen H, Moldawer L, Nathan C, Lowry S, Cerami A (1988) Macrophages secrete a novel heparin-binding protein with inflammatory and neutrophil chemokinetic properties. J Exp Med 172:911–919

10 Protein Modifications and Interactions in Alzheimer's Disease

M.A. Smith and G. Perry

10.1 Introduction

Alzheimer's disease, an age-related neurodegenerative disease, is characterized by several coincidental features including neuronal loss, neurofibrillary tangles and senile plaques. While the precise mechanism(s) underlying Alzheimer's disease are incompletely understood, it is apparent that age is essential since the disease rarely strikes prior to age 55. Many other conditions such as emphysema, arthritis, and atherosclerosis also show an age-related penetrance and occur after the normal reproductive life span. This latter aspect is especially important since there is very little evolutionary pressure to maintain normal physiological relationships between biomolecules and preclude the development of these diseases. This lack of evolutionary selection allows altered or novel macromolecular relationships. Our hypothesis is that altered protein interactions and protein modifications could destabilize homeostatic relationships and/or be themselves damaging to the cellular system

promoting the pathogenesis of Alzheimer's disease and other chronic
degenerative diseases.

10.2 Protein Interactions

Protein interactions are essential to the homeostatic regulation of multi-
cellular organisms by defining macromolecular assemblies which,
together with the primary protein sequence, are key determinants of
conformation. These aspects are of particular relevance in Alzheimer's
disease, where senile plaques and neurofibrillary tangles, abnormal
pathological structures not normally found in the brain, are polymeric
macromolecular assemblies of proteins arranged in a β-sheet conformation. We suspected that β-sheet fibrils arise through the formation of
novel interactions, perhaps involving an altered protein conformation.

Neurofibrillary tangles and senile plaques are compositionally complex although the main components, microtubule-associated protein τ
and amyloid-β, respectively, are well characterized. The obligate coin-
cidence of neurofibrillary tangles and senile plaques in Alzheimer's
disease led us to speculate that either there was a direct biochemical
relationship between the lesions and that this relationship was mediated
through protein interactions (Perry 1993) and/or the lesions are formed
by a common pathogenic mechanism (Smith et al. 1995a).

Since the normal physiological role of τ is to bind to and stabilize a
number of different protein polymers, τ could perform a similar, but
novel, role in the diseased brain (Smith et al. 1995b). Using an *in situ*
binding assay, we found that nanomolar concentrations of τ specifically
interact with a saturable site found in all senile plaques. To map the site
of interaction between τ and senile plaques we used (1) antibodies to
sterically hinder interaction and (2) preincubation of τ with protein to
adsorb consequent interaction and found that while antibodies/peptides
to amyloid-β were unable to affect binding, antibodies/peptides to the β
-protein precursor (βPP) were able to completely inhibit binding. These
data suggest that the key element in τ interaction with senile plaques is
βPP and further characterization, using an array of overlapping βPP
peptides and specific antibodies, demonstrated that $\beta PP_{714-723}$ is essential for high affinity binding (Fig. 1; using the numbering of βPP_{770};
Kitaguchi et al. 1988). τ binds in a saturable manner with a K_D of

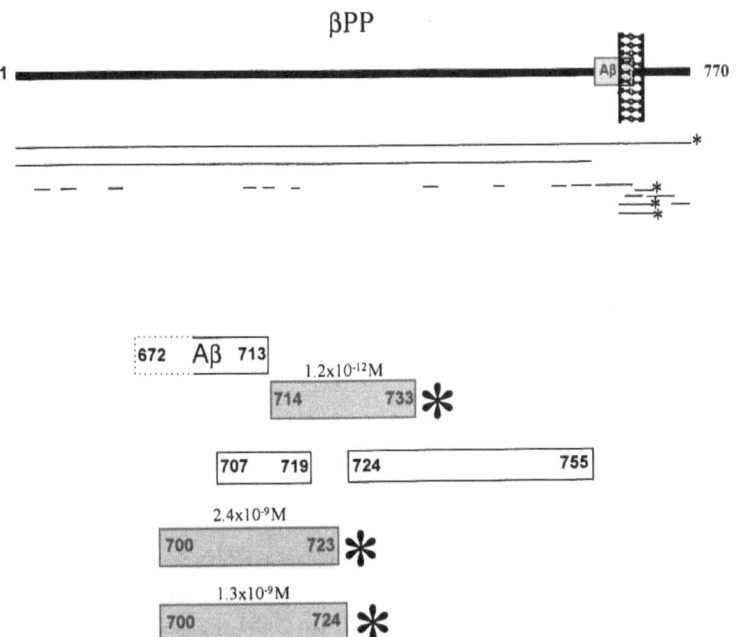

Fig. 1. The binding of τ to senile plaques was completely inhibited by βPP and by a number of competing peptides (* and *shaded*) as indicated below the βPP sequence. Concentrations given are those required to completely block τ binding to senile plaques. The putative minimal τ-binding domain of βPP contains βPP714-723. Numbering system of βPP770

9.6 n*M* and, significantly, the interaction of τ with βPP (and senile plaques) is β-sheet conformation-dependent since τ binding is blocked by either stabilization of α-helices or incubation with Congo red, a dye that avidly binds to β-sheets (Smith et al. 1995b). It seems more than coincidental that the one element common to all amyloids, i.e., β-sheet conformation (Glenner 1980), is necessary for the interaction of τ and βPP. Another interesting aspect regarding the τ-βPP interaction is that the τ-binding domain spans a mutation site associated with familial forms of Alzheimer's disease, i.e., βPP717 (Goate et al. 1991), and one hypothesis is that mutations at this site alter the interaction with τ in a manner that induces disease development.

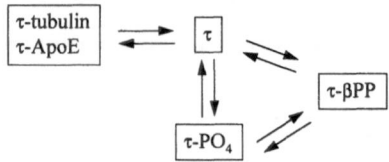

Fig. 2. Multiple interactions of τ with other proteins implicated in the pathogenesis of Alzheimer's disease. The simplified scheme requires further studies to determine if binding sites overlap. Phosphorylation frees τ from interaction with apolipoprotein E (ApoE) and tubulin, thereby increasing, by mass action, the interaction of τ with βPP

Unlike τ interaction with other cytoskeletal proteins such as microtubules and actin (Lindwall and Cole 1984; Selden and Pollard 1983) the requirements for the interaction of τ with βPP are not mediated by the τ phosphorylation state and, in addition, τ isoforms from human, bovine, or mouse brain all bind to βPP (Smith et al. 1995b). Moreover, since βPP has no effect on microtubule assembly, these findings suggest that the βPP-binding domain of τ is distinct from the tubulin/microtubule-binding domain.

In Alzheimer's disease, where τ phosphorylation (Grundke-Iqbal et al. 1986) likely affects microtubule assembly (Gustke et al. 1992), the interaction of τ with βPP might be of particular relevance. Although not directly affecting the interaction of τ and βPP, increased phosphorylation of τ might serve as a "pathological switch" by both decreasing the interaction of τ with tubulin, thereby creating cytoskeletal disturbances characteristic of Alzheimer's disease (Ellisman et al. 1987; Perry et al. 1991), and by increasing the availability of τ for interaction with βPP. Importantly, and paralleling the known function of τ to promote fibrils with tubulin (Connolly et al. 1977), actin (Selden and Pollard 1983) and neurofilament (Miyata et al. 1986), τ/βPP interaction may promote fibrillogenesis (Giaccone et al. 1996). Therefore, the effects of phospho-

Fig. 3A-C. Immunoelectron microscopy of rat brain demonstrates that βPP is ▶ associated with **A** membranes of synaptic vesicles (*arrowheads*); **B** outer mitochondrial membranes (*arrowheads*); and **C** some microtubules (*arrowheads*) but not all (*arrow*). Immunoperoxidase. *Scale bars* **A**: 2 μm; **B**: 2.5 μm; **C**: 1 μm

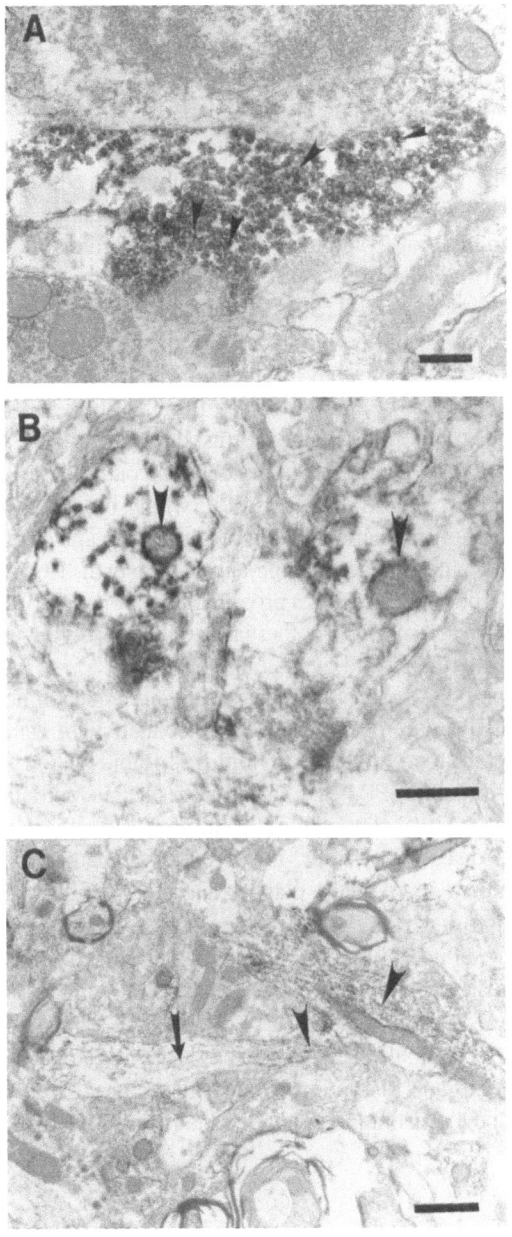

rylation, although lowering by tenfold the self-assembly of τ into neuro-fibrillary tangles (Wischik et al. 1995), would be to free a pool of τ from microtubule interaction, promoting βPP binding and in turn inducing τ-βPP polymer formation by nucleating and promoting fibril formation (Fig. 2).

Obviously, while a simple model of τ/tubulin or τ/βPP interaction is attractive (Fig. 2), one must also consider the effects of other (competing) interactions. For example, apolipoprotein E, a component of both neurofibrillary tangles and senile plaques, interacts with both dephosphorylated τ and amyloid-β (Strittmatter et al. 1993; Richey et al. 1995). The interaction of apolipoprotein E with dephosphorylated τ, by preventing phosphorylation, may promote microtubule assembly (Fig. 2; Strittmatter et al. 1994) and it will be of interest to determine whether the different apolipoprotein E isoforms show differential binding affinities for τ. Another protein interaction of pathological importance might be that between trypsin and βPP, which by analogy with other protease/inhibitor interactions might result in cleavage of βPP to generate potentially amyloidogenic fragments (Smith et al. 1995c). Additionally, this is another example of a protein interaction with βPP and could, as in Fig. 2, affect a myriad of other interactions involving either trypsin or βPP or other related and interacting elements.

The scenario set up in Fig. 2 is based upon the idea of novel interactions. Normally, apolipoprotein E is associated with astrocytes, τ is associated with microtubules, amyloid-β is found at low levels in the CSF and serum, and βPP is associated with plasma, cytoplasmic membranes and the cytoskeleton (Fig. 3; Autilio-Gambetti et al. 1988). In Alzheimer's disease, however, all of these molecules together form integral components of senile plaques and neurofibrillary tangles. It is our contention that while some of the interactions responsible for the development of neurofibrillary tangles and senile plaques are novel, many others are seen physiologically, but in Alzheimer's disease there is an alteration in this interaction. For example, βPP is associated with a subclass of microtubules (Fig. 3), is released as the full-length protein, and with membrane disruption (Praprotnik et al. 1996a) can interact with intra- and extracellular compartments (Praprotnik et al. 1996b).

10.3 Oxidative Stress

Ever since the Cambrian era, organisms benefiting from the greater energy production offered by aerobic metabolism have found it necessary to deal with the problem of oxygen reactivity with biomolecules. To this end, enzymes such as superoxide dismutase, catalase, and peroxidase and antioxidants such as ascorbate, tocopherol, and glutathione are necessary to protect against the potentially devastating effects of uncontrolled oxidative stress. Nevertheless, oxidative damage is seen *in vivo* and one theory of aging is that accumulation of oxidative damage underlies aging (Harman 1956).

One of the identified categories of oxidative modification of proteins, the Maillard reaction, is the nonenzymatic reaction between the carbonyl group of reducing sugars and the amino groups of protein. Sugar addition through Schiff base chemistry and subsequent reorganization generates numerous products (Fig. 4) and as well as creating an irreversible modification that also creates, through redox cycling, an array of potent oxidants including $O_2\bullet$ and OH$\bullet$ (Hunt and Wolff 1991; Yan et al. 1994, 1995). While the bulk of Maillard rearrangements are uncharacterized, antibodies to two well-defined products, pentosidine and pyrraline, have been produced (Miyata and Monnier 1992; Smith et al. 1994a) and, in tissue from Alzheimer's disease, recognized neurofibrillary tangles and senile plaques (Smith et al. 1994a). These findings parallel those indicating that amyloid-β fibrillogenesis is increased by AGE (advanced glycation end products; Vitek et al. 1994), that the τ component of neurofibrillary tangles contains AGE modifications (Yan et al. 1994), and that AGE-modified τ induces oxidant stress in neuroblastoma cells (Yan et al. 1995) and clearly establish AGE products as an important facet of the pathology of Alzheimer's disease.

The synergistic role between AGE and oxidative stress (Smith et al. 1995a) led us to look for evidence of oxidative stress in Alzheimer's disease. Heme oxygenase-1 (HO-1), an inducible enzyme isoform that is involved in the first catalytic step in converting the prooxidant heme to the antioxidant bilirubin, is induced during periods of oxidative stress. Reverse transcription-polymerase chain reaction (RT-PCR) amplification of HO-1 mRNA showed a robust induction in Alzheimer's disease over age-matched control paralleling the brain regions susceptible to pathology (Premkumar et al. 1995). Immunocytochemical ana-

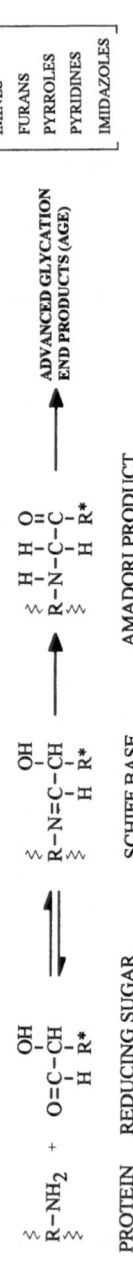

Fig. 4. The initial reaction between a reducing sugar or other carbonyl compound with the amino group of a protein generates an equilibrium Schiff base which through rearrangement forms a stable Amadori product. Through further rearrangements, fragmentations and condensations form a wide variety of advanced glycation end products (AGEs) including imines, furans, pyrroles, pyridines, and imidazoles

lyses confirmed these results and showed that HO-1 is increased in association with nearly all of the neurofibrillary pathologies, including senile plaque neurites, neurofibrillary tangles, and neuropil threads (Smith et al. 1994b). These findings, together with previous reports of increased Cu-Zn superoxide dismutase immunoreactivity in neurofibrillary tangles (Pappolla et al. 1993), suggest neuronal response to oxidant insult in Alzheimer's disease.

In addition to τ, neurofilaments in Alzheimer's disease show extensive oxidative modification (Smith et al. 1995d). A subclass of monoclonal antibodies raised to neurofibrillary tangles recognizes neurofilament proteins on immunoblots derived from neurofibrillary tangles but not neurofilaments from control cases. These antibodies are carbonyl-dependent, AGE being a category of carbonyl modification, and recognize neurofilaments on tissue sections and heavy neurofilament subunits on immunoblots only following carbonyl treatment, while they recognized the neurofilament component of neurofibrillary tangles in Alzheimer's disease without prior treatment (Smith et al. 1995d).

Neurofilament proteins have properties that make them ideal for carbonyl attack in that they have the longest transit time in the axon and contain abundant lysines (Shaw 1991). Since neurofilament transport is of a single macromolecular assembly (Lasek et al. 1983), from the cell body to terminal, it is tempting to speculate that axonal neurofilament subunits are held in those assemblies by cross-links involving the initially reversible carbonyl modifications (Schiff-base type) distinct from the more advanced products recognized by the antibodies to neurofibrillary tangles or AGE, and that disregulation of the extent and type of oxidative modification plays a role in cytoskeletal alteration in AD and other neurodegenerative diseases. There are two lines of investigation supporting this view. First, the cytoskeleton-derived filamentous inclusions in Parkinson (Galloway et al. 1992) and Pick (Perry et al. 1987) diseases as well as progressive supranuclear palsy (Tabaton et al. 1988) are insoluble in denaturants, suggesting they, like the filaments of Alzheimer's disease, are crosslinked. Second, in motor neuron disease, mutations in Cu/Zn superoxide dismutase are associated with neurofilament accumulation and AGE modification of axonal protein (Chou, Smith and Perry, unpublished observations). Moreover, following carbonyl intoxication there are well known alterations of neurofilaments (Sayre et al. 1985).

The molecular organization of neurofilament protein suggests an amino acid sequence arrangement that might control carbonyl bonds. In the two larger neurofilament subunits, C-terminal "tails" contain numerous repeats of lysine followed by serine and proline (KSP), with serine often phosphorylated (Shaw 1991). The function of phosphorylation is unknown (Sternberger and Sternberger 1983), but it occurs as neurofilaments enter axons and may play a role in subunit association with preformed axonal neurofilaments (Nixon 1991). One aspect of AGE chemistry is the need for close juxtaposition of adjacent carbonyl modifications for Maillard reorganization. Phosphates, through charge repulsion, may minimize the formation of these bonds, thus maintaining the cytoskeleton's supramolecular structure but not creating irreversible crosslinks. Imbalance of this system in oxidative stress or abnormal phosphorylation may underlie cytoskeleton susceptibility to disruption in superoxide dismutase (SOD) mutants (Dal Canto and Gurney 1994), carbonyl intoxication (Sayre et al. 1985), or hyperphosphorylation (Trojanowski et al. 1993)

10.4 Conclusion

Understanding how normal homeostatic interactions can be maintained might provide novel strategies to lower the prevalence of Alzheimer's disease and other age-related degenerative diseases.

Acknowledgments. This work represents studies supported by grants from the National Institutes of Health, the American Health Assistance Foundation, the American Federation of Aging, and the American Philosophical Society.

References

Autilio-Gambetti L, Morandi A, Tabaton M, Schaetzle B, Kovacs D, Perry G, Greenberg B, Gambetti P (1988) The amyloid precursor protein of Alzheimer disease is expressed as a 130 kDa polypeptide in various cultured cell types. FEBS Lett 241:94–98
Connolly JA, Kalnins VI, Cleveland DW, Kirschner MW (1977) Immunofluorescent staining of cytoplasmic and spindle microtubules in mouse fibroblasts with antibody to tau protein. Proc Natl Acad Sci USA 74:2437–2440

Dal Canto MC, Gurney ME (1994) Development of central nervous system pathology in a murine transgenic model of human amyotrophic lateral sclerosis. Am J Pathol 145:1271–1279

Ellisman M, Ranganathan R, Deerinck T, Young S, Terry R, Mirra S (1987) Neuronal fibrillar cytoskeleton and endomembrane system organization in Alzheimer's disease. In: Perry G (ed) Alteration in the neuronal cytoskeleton in Alzheimer's disease. Plenum, New York, pp 61–73

Galloway PG, Mulvihill P, Perry G (1992) Filaments of Lewy bodies contain insoluble cytoskeletal elements. Am J Pathol 140:809–822

Giaccone G, Pedrotti B, Migheli A, Verga L, Perez J, Racagni G, Smith MA, Perry G, De Gioia L, Selvaggini C, Salmona M, Ghiso J, Frangione B, Islam K, Bugiani O, Tagliavini F (1996) βPP and tau interaction: a possible link between amyloid and neurofibrillary tangles in Alzheimer's disease. Am J Pathol 148:79–87

Glenner GG (1980) Amyloid deposits and amyloidosis. N Engl J Med 302:1283–1292

Goate A, Chartier-Harlin M-C, Mullan M, Brown J, Crawford F, Fidani L, Guiffra L, Haynes A, Irving N, James L, Mant R, Newton P, Rooke K, Roques P ,Talbot C, Pericak-Vance M, Roses A, Williamson R, Rossor M, Owen M, Hardy J (1991) Segregation of a missense mutation in the amyloid precursor protein gene with familial Alzheimer's disease. Nature 349:704–706

Grundke-Iqbal I, Iqbal K, Tung YC, Quinlan M, Wisniewski HM, Binder LI (1986) Abnormal phosphorylation of the microtubule-associated protein tau in Alzheimer cytoskeletal pathology. Proc Natl Acad Sci USA 83:4913–4917

Gustke N, Steiner B, Mandelkow EM, Biernat J, Meyer HE, Goedert M, Mandelkow E (1992) The Alzheimer-like phosphorylation of tau protein reduces microtubule binding and involves Ser-Pro and Thr-Pro motifs. FEBS Lett 307:199–205

Harman D (1956) Aging: a theory based on free radical and radiation chemistry. J Gerontol 11:298–300

Hunt JV, Wolff SP (1991) Oxidative glycation and free radical production: a causal mechanism of diabetic complications. Free Radic Res Commun 12-13:115–123

Kitaguchi N, Takahashi Y, Tokushima Y, Shiojiri S, Ito H (1988) Novel precursor of Alzheimer's disease amyloid protein shows protease inhibitory activity. Nature 331:530–532

Lasek RJ, McQuarrie IG, Brady ST (1983) Transport of cytoskeletal and soluble proteins in neurons. In: Opiatka A, Balaban M (eds) Biological structures and coupled flows. Academic, New York, pp 329–347

Lindwall G, Cole RD (1984) Phosphorylation affects the ability of tau protein to promote microtubule assembly. J Biol Chem 259:5301–5305

Miyata S, Monnier VM (1992) Immunocytochemical detection of advanced glycosylation end products in diabetic tissues using monoclonal antibody to pyrraline. J Clin Invest 89:1102–1112

Miyata Y, Hoshi M, Nishida E, Minami Y, Sakai H (1986) Binding of microtubule-associated protein 2 and tau to the intermediate filament reassembled from neurofilament 70-kDa subunit protein. Its regulation by calmodulin. J Biol Chem 261:13026–13030

Nixon RA (1991) Axonal transport of cytoskeletal proteins. In: Burgoyne RD (ed) The neuronal cytoskeleton. Wiley-Liss, New York, pp 283–307

Pappola MA, Omar RA, Kim KS, Robakis NK (1993) Immunohistochemical evidence of antioxidant stress in Alzheimer's disease. Am J Pathol 140:621–628

Perry G (1993) Neuritic plaques in Alzheimer disease originate from neurofibrillary tangles. Med Hypotheses 40:257–258

Perry G, Stewart D, Friedman R, Manetto V, Autilio-Gambetti L, Gambetti P (1987) Filaments of Pick's bodies contain altered cytoskeleton elements. Am J Pathol 127:559–568

Perry G, Kawai M, Tabaton M, Onorato M, Mulvihill P, Richey P, Morandi A, Connolly J, Gambetti P (1991) Neuropil threads of Alzheimer's disease show a marked alteration of the normal cytoskeleton. J Neurosci 11:1748–1755

Praprotnik D, Smith MA, Richey PL, Vinters HV, Perry G (1996a) Plasma membrane fragility in dystrophic neurites in senile plaques of Alzheimer's disease: an index of oxidative stress. Acta Neuropathol (Berl) (in press)

Praprotnik D, Smith MA, Richey PL, Vinters HV, Perry G (1996b) Filament heterogeneity within the dystrophic neurites of senile plaques suggests blockage of fast axonal transport in Alzheimer's disease. Acta Neuropathol (Berl) 91:1–5

Premkumar DRD, Smith MA, Richey PL, Petersen RB, Castellani R, Kutty RK, Wiggert B, Perry G, Kalaria RN (1995) Induction of heme oxygenase-1 mRNA and protein in neurocortex and cerebral vessels in Alzheimer's disease. J Neurochem 65:1399–1402

Richey PL, Siedlak SL, Smith MA, Perry G (1995) Apolipoprotein E interaction with the neurofibrillary tangles and senile plaques in Alzheimer disease: implications for disease pathogenesis. Biochem Biophys Res Commun 208:657–663

Sayre LM, Autilio-Gambetti L, Gambetti P (1985) Pathogenesis of experimental giant neurofilamentous axonopathies: a unified hypothesis based on chemical modification of neurofilaments. Brain Res Rev 357:69–83

Selden SC, Pollard TD (1983) Phosphorylation of microtubule-associated proteins regulates their interaction with actin filaments. J Biol Chem 258:7064–7071

Shaw G (1991) Neurofilament proteins. In: Burgoyne RD (ed) The neuronal cytoskeleton. Wiley-Liss, New York, pp 185–214

Smith MA, Taneda S, Richey PL, Miyata S, Yan S-D, Stern D, Sayre LM, Monnier VM, Perry G (1994a) Advanced Maillard reaction end products are associated with Alzheimer's disease pathology. Proc Natl Acad Sci USA 91:5710–5714

Smith MA, Kutty RK, Richey PL, Yan S-D, Stern D, Chader GJ, Wiggert B, Petersen RB, Perry G (1994b) Heme oxygenase-1 is associated with the neurofibrillary pathology of Alzheimer's disease. Am J Pathol 145:42–47

Smith MA, Sayre LM, Monnier VM, Perry G (1995a) Radical AGEing in Alzheimer's disease. Trends Neurosci 18:172–176

Smith MA, Siedlak SL, Richey PL, Mulvihill P, Ghiso J, Frangione B, Tagliavini F, Giaccone G, Bugiani O, Praprotnik D, Kalaria RN, Perry G (1995b) Tau protein directly interacts with the amyloid β-protein precursor: implications for Alzheimer's disease. Nature Med 1:365–369

Smith MA, Dunbar CE, Miller EJ, Perry G (1996) Trypsin interaction with senile plaques of Alzheimer disease is mediated by β-protein precursor. Mol Chem Neuropathol 27:145–154

Smith MA, Rudnicka-Nawrot M, Richey PL, Praprotnik D, Mulvihill P, Miller CA, Sayre LM, Perry G (1995d) Carbonyl-related posttranslational modification of neurofilament protein in the neurofibrillary pathology of Alzheimer's disease. J Neurochem 64:2660–2666

Sternberger LA, Sternberger NH (1983) Monoclonal antibodies distinguish phosphorylated and nonphosphorylated forms of neurofilaments in situ. Proc Natl Acad Sci USA 80:6126–6130

Strittmatter WJ, Weisgraber KH, Huang DY, Dong LM, Salvesen GS, Pericak-Vance M, Schmechel D, Saunders AM, Goldgaber D, Roses AD (1993) Binding of human apolipoprotein E to synthetic amyloid beta peptide: isoform-specific effects and implications for late-onset Alzheimer disease. Proc Natl Acad Sci USA 90:8098–8102

Strittmatter WJ, Weisgraber KH, Goedert M, Saunders AM, Huang D, Corder EH, Dong LM, Jakes R, Alberts MJ, Gilbert JR, Han S-H, Hulette C, Einstein G, Schmechel DE, Pericak-Vance MA, Roses AD (1994) Hypothesis: microtubule instability and paired helical filament formation in Alzheimer disease brain are related to apolipoprotein E genotype. Exp Neurol 125:163–171

Tabaton M, Whitehouse PJ, Perry G, Davies P, Autillio-Gambetti L, Gambetti P (1988) Alz 50 recognizes abnormal filaments in Alzheimer's disease and progressive supranuclear palsy. Ann Neurol 24:407–413

Trojanowski JQ, Schmidt ML, Shin R-W, Bramblett GT, Goedert M, Lee VM-
    Y (1993) PHFτ(A68): from pathological marker to potential mediator of
    neuronal dysfunction and degeneration in Alzheimer's disease. Clin Neuro-
    sci 1:184–191
Vitek MP, Bhattacharya K, Glendening JM, Stopa E, Vlassara H, Bucala R,
    Manogue K, Cerami A (1994) Advanced glycation end products contribute
    to amyloidosis in Alzheimer disease. Proc Natl Acad Sci USA 91:4766–
    4770
Wischik CM, Edwards PC, Harrington CR, Mukaetova-Ladinska E, Roth M
    (1995) A clinico-pathological view of molecular pathophysiology and pro-
    phylaxis of AD. Alzheimer Res 1 [Suppl 1]:12 (abstract)
Yan SD, Chen X, Schmidt AM, Brett J, Godman G, Zou YS, Scott CW, Ca-
    puto C, Frappier T, Smith MA, Perry G, Yen SH, Stern D (1994) Glycated
    tau protein in Alzheimer disease: a mechanism for induction of oxidant
    stress. Proc Natl Acad Sci USA 91:7787–7791
Yan SD, Yan SF, Chen X, Fu J, Chen M, Kuppusamy P, Smith MA, Perry G,
    Godman GC, Nawroth P, Zweier JL, Stern D (1995) Non-enzymatically
    glycated tau in Alzheimer's disease induces neuronal oxidant stress result-
    ing in cytokine gene expression and release of amyloid-β peptide. Nature
    Med 1:693–699

11 Protein Aging and Its Relevance to the Etiology of Alzheimer's Disease

E.R. Stadtman

11.1 Reactive Oxygen Species that Modify Proteins

The oxidative modification of proteins is elicited by a number of reactive oxygen species (ROS) that are produced endogenously as minor by-products of normal electron transport processes in metabolism, by exposure to oxidative stress as a consequence of metabolic disorders or environmental challenges, and by metal-catalyzed oxidation (MCO) systems. The ROS most commonly implicated in protein modification are hydroxyl radical (OH$^\bullet$), superoxide anion radical (O$_2$$^{\bullet-}$), nitric oxide radical (NO$^\bullet$), thiyl radical (RS$^\bullet$), perferryl radical (FeO$_2$$^{2+}$), ferryl radical (FeO^{2+}), and various non-radical species, such as hydrogen peroxide (H$_2$O$_2$), ozone (O$_3$), singlet oxygen (^{1}O$_2$), peroxynitrite (HONOO), nitronium ion (NO$_2$$^+$), and hypochlorous acid (HOCl). Whereas each of these ROS may contribute to biological damage under a particular set of conditions, it is generally accepted that OH$^\bullet$ is the most damaging ROS under most physiological conditions. Hydroxyl radicals are readily generated by at least four different mechanisms: (a) the homolytic cleavage of water by ionizing radiation (X-rays, gamma rays), reaction (1), which describes the overall reaction in which aqueous electrons (e$_{aq}$$^-$) , H$_2O^+$, and an excited state of water (H$_2$O*) are intermediates (Swallow 1960); (b) cleavage of H$_2$O$_2$ by reaction with either Fe(II) or Cu(I), the Fenton reaction, reaction (2); (c) by homolytic cleavage of peroxynitrite, reaction (3) (Beckman et al. 1990); (d) by reaction of ozone with a phenol (PH), reaction (4) (Grimes et al. 1988; Pryor 1994).

$$H_2O \rightarrow H^\bullet + OH^\bullet \ (1)$$
$$H_2O_2 + Fe(II) \rightarrow OH^\bullet + OH^- + Fe(III) \ (2)$$
$$HONOO \rightarrow OH^\bullet + NO_2 \ (3)$$
$$PH + O_3 \rightarrow OH^\bullet + P + O_2 \ (4)$$

Except for individuals exposed accidentally or therapeutically to high doses of radiation, or to high levels of ozone in the atmosphere, the principal source of OH$^\bullet$ in biological systems is via the metal-catalyzed reaction (2). The production of OH$^\bullet$ by MCO systems is dependent upon the availability of H$_2$O$_2$, and of Fe(II) or Cu(I), whose concentrations in vivo are tightly controlled at very low levels.

11.1.1 Hydrogen Peroxide Production

H$_2$O$_2$ is a normal end product of reactions catalyzed by various oxidases (Brunori and Rotilio 1984). It is formed also by the dismutation of O$_2$$^{\bullet-}$, reaction (5), which is unavoidably formed by autooxidation of the reduced forms of flavoproteins and quinones during the transfer of electrons to cytochrome C in terminal respiration (Bovaris et al. 1972; Halliwell and Gutteridge 1989).

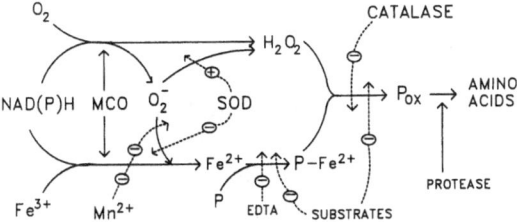

Fig. 1. Iron-catalyzed generation of reactive oxygen species. *MCO*, metal-catalyzed oxidation systems; *P*, protein; *SOD*, superoxide dismutase

$$2O_2^{\bullet-} + 2H^+ \rightarrow H_2O_2 + O_2 \quad (5)$$

As discussed previously (Stadtman 1988a,b, 1995), $O_2^{\bullet-}$ and H_2O_2 can be formed also by members of a class of flavoprotein dehydrogenases that under normal conditions catalyze oxidation-reduction reactions in intermediary metabolism, but, when the substrate levels of electron acceptors become rate-limiting, the reduced forms of these enzymes may react with O_2 to form $O_2^{\bullet-}$ and H_2O_2.

11.1.2 Metal-Catalyzed Oxidation Systems

A number of different enzymatic MCO systems have been shown to catalyze the modification (inactivation) of enzymes (reviewed by Stadtman 1990). These include reactions catalyzed by xanthine oxidase, glucose oxidase, horseradish peroxidase (Stadtman and Wittenberger 1985), NADH/NADPH oxidases and cytochrome P450 oxido-reductases (Oliver et al. 1982; Fucci et al. 1983), and reactions catalyzed by a number of nonenzymatic MCO systems using ascorbate (Levine 1983; Chevion 1988) or sulfhydryl compounds (Kim et al. 1985) as electron donors. All of these MCO systems have been shown to catalyze the reduction of Fe(III) to Fe(II) and the reduction of O_2 to H_2O_2 by $O_2^{\bullet-}$-dependent or $O_2^{\bullet-}$-independent pathways (Fig. 1).

11.1.2.1 Availability of Iron

Although iron is present in substantial amounts in most tissues, it exists almost exclusively in the form of protein complexes (ferritin, transferrin, hemosiderin, lactoferrin, hemogloblin, cytochromes, protein-bound non-heme iron-sulfur clusters), and is generally unavailable for Fenton-type chemistry (Halliwell and Gutteridge 1989). Nevertheless, under conditions of oxidative stress, mechanical injury, or during the degradation of heme proteins, iron is released from these stores and becomes available for Fenton chemistry (Halliwell and Gutteridge 1989; Cairo et al. 1995; Allen et al. 1994; Lamb and Leake 1994).

11.2 Protein Modification Reactions

11.2.1 Site-Specific Modification of Proteins

In the presence of ionizing radiation (Swallow 1960; Garrison et al. 1962; Garrison 1987) or high concentrations of metal ions [Fe(III), Cu(II)], hydrogen peroxide, ascorbate, and physiological chelating agents, such as citrate or nucleoside di- and triphosphates, hydroxyl radicals are formed in the bulk solution and will react randomly with almost any biomolecule (protein, nucleic acid, lipid) in close proximity to the site of radical generation (Huggins et al. 1993; Neuzil et al. 1993). However, there is considerable evidence that in the presence of physiological concentrations of metal ions, hydrogen peroxide, and an electron donor, the modification of proteins is a site-specific process that involves the binding of metal ion to metal binding sites on the protein, followed by reaction with H_2O_2 to form highly reactive radical species (OH$^\bullet$, ferryl, perferryl) that react preferentially with amino acid residues at the metal binding site (Fucci et al. 1983; Stadtman 1986; Chevion 1988; Stadtman and Oliver 1991).

11.2.2 Oxidation of Amino Acid Side Chains

Among other modifications: histidine residues are converted to 2-oxo-histidine residues (Uchida and Kawakishi 1993, 1994) and/or to asparagine/aspartic acid residues (Creeth et al. 1983; Farber and Levine 1986); arginine and proline residues are converted to glutamic semialdehyde residues (Amici et al. 1989; Climent et al. 1989); proline residues are converted to pyroglutamyl residues (Creeth et al. 1983;

Amici et al. 1989) and to *cis/trans*-4-hydroxyproline residues (Poston 1988), and to 2-pyrrolidine derivatives with concomitant peptide bond cleavage (Uchida et al. 1990); cysteine residues are converted to disulfides (Alonso et al. 1992; Takahashi and Goto 1990; Zhou and Gafni 1991); threonine residues are converted to 2-amino-3-ketobutyric acid residues (Taborsky 1973); methionine residues are converted to methionine sulfoxide residues (Vogt 1995). In addition, with prolonged exposure to MCO systems, some proteins undergo aggregation (Kim et al. 1985), especially in the case of neurofilaments (Troncoso et al. 1995).

It is noteworthy that phenylalanine, tyrosine and tryptophan, valine, and leucine residues are not generally modified by MCO systems in the presence of physiological concentrations of Fe(III) and H_2O_2, presumably because these amino acids are not commonly at metal binding sites on proteins. In contrast, when proteins are exposed to ionizing radiations, the aromatic and hydrophobic amino acid residues as well as histidine and the sulfur-containing amino acid residues are clearly the preferred targets (Swallow 1960; Garrison 1987). The sulfur-containing amino acid residues, histidine, and the aromatic amino acid residues are also primary targets of ozone (Pryor et al. 1994; Berlett et al. 1991). The aromatic and sulfur-containing amino acids are also primary targets of peroxynitrite (Berlett et al. 1995; Ischiropoulos and Al-Mehdi 1995).

11.2.3 Peptide Bond Cleavage

Free radical-mediated peptide bond cleavage occurs by at least four different mechanisms (Garrison 1987). As illustrated in Fig. 2, these include: (a) cleavage by the α-amidation pathway in which the N-terminal amino acid of one protein fragment is present as an α-ketoacyl derivative and the C-terminal amino acid of the other fragment is present as the amide derivative; (b) cleavage by the diamide pathway, which leads to formation of a diamide derivative of the C-terminal end of one peptide fragment and transient formation of the isocyanate derivative of the N-terminal amino acid of the other fragment; (c) cleavage via oxidation of glutamyl side chains leading to a pyruvyl derivative of the N-terminal amino acid residue of one peptide fragment; (d) oxidation of prolyl residues to a pyrrolidone derivative and concomitant

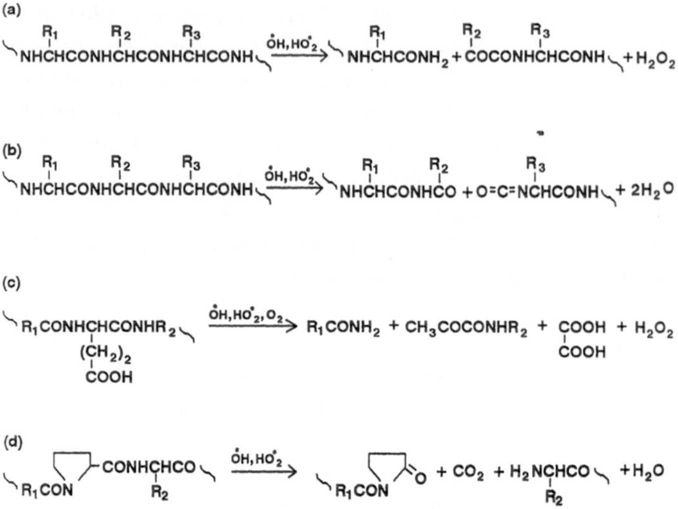

(a)

(b)

(c)

(d)

Fig. 2a-d. Peptide bond cleavage reactions. Products formed by: **a** the α-amidation pathway; **b** the diamide pathway; **c** the glutamate oxidation pathway; **d** the proline oxidation pathway

peptide bond cleavage as shown by Uchida et al. (1990). It is noteworthy that peptide bond cleavage reactions by the α-amidation and glutamic acid oxidation pathways lead to the formation of carbonyl derivatives.

11.2.4 Indirect Formation of Protein Carbonyl Derivatives

The generation of protein carbonyl groups is not restricted to direct oxidation of amino acid side chains or peptide bond cleavage reactions. The introduction of carbonyl groups into proteins occurs also by reactions with oxidation products of polyunsaturated fatty acids and with reducing sugars and products derived therefrom by metal-catalyzed reactions.

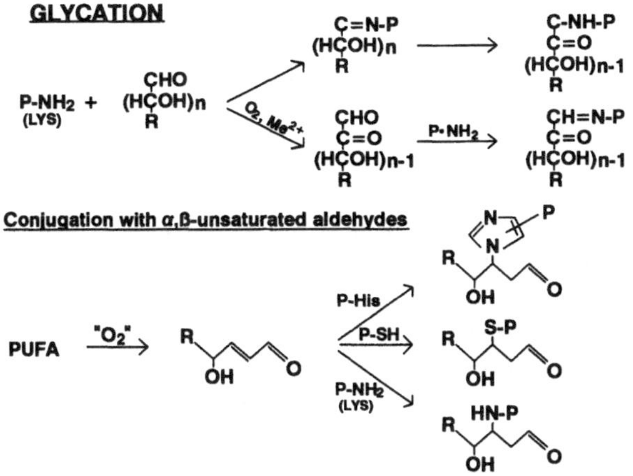

Fig. 3. Protein carbonyls produced by reactions with carbohydrates (glycation) or by reaction with αβ-unsaturated aldehydes. *P*, protein

11.2.4.1 Modification by Lipid Peroxidation Products

Peroxidation of polyunsaturated fatty acids leads to the production of several aldehydes, including malondialdehyde and the highly cytotoxic 4-hydroxy-2-nonenal (HNE) (Esterbauer et al. 1991). As illustrated in Fig. 3, HNE can undergo Michael-type addition reactions with protein sulfhydryl groups (Schauenstein and Esterbauer 1979), and with lysine ε amino groups and histidine residues of proteins (Uchida and Stadtman 1993). In each case, the aldehyde moiety of the HNE is preserved and therefore can contribute to the protein carbonyl pool, or it may undergo secondary reactions with ε amino groups of other lysine residues to form intra- or intermolecular cross-linked derivatives (Uchida and Stadtman 1993; Friguet et al. 1994). It is worth noting that such HNE cross-linked proteins are resistant to degradation by the multicatalytic protease and, in fact, they inhibit the ability of the protease to degrade the oxidized forms of other proteins (Friguet et al. 1994). The physiological significance of HNE-protein conjugation is indicated by the demonstration that antibodies raised against HNE-treated, low density lipoprotein (LDL) react with oxidized forms of LDL (Palinski et al. 1989) and also by the demonstration that HNE-conjugates of apolipoprotein B-100 are formed when LDL is exposed to HNE or to oxidation by Cu(II) (Uchida et al. 1994). In addition, it was demonstrated by immunochemical analysis that HNE-protein conjugates are present in the renal proximal tubules of rats following treatment with the renal carcinogen, ferric nitrilotriacetate (Toyokuni et al. 1994).

11.2.4.2 Glycation and Glycoxidation

Carbonyl groups can also be introduced by reactions with reducing sugars or with carbohydrate oxidation products in a process referred to as glycation. As shown in Fig. 3, glycation involves reaction of the carbonyl groups of reducing sugars with the N-terminal amino group or the ε-amino group of lysine residues of proteins to form Schiff base adducts that undergo subsequent Amadori rearrangements to yield ketoamines. The Amadori products are highly sensitive to metal-catalyzed reactions (glycoxidation) leading to either N^ϵ-carboxymethyl-lysine residues of the protein or to reactions with arginine residues to form pentosidine protein cross-linked derivatives (Monnier 1990; Monnier et al. 1995). Alternatively, the Amadori products may undergo dehydration/oxidative fragmentation to form deoxyosones that undergo further reactions to yield highly fluorescent cross-linked derivatives of poorly defined structures (Monnier 1990). These are collectively referred to as Maillard products or advanced glycosylation end products (AGEs). These Maillard products accumulate during aging and are probably implicated in diabetes and some eye diseases (Monnier 1990; Wells-Knecht et al. 1993). It has been reported that some of the Maillard products can generate free radicals in the absence of metal ions and might be implicated in atherogenesis (Mullarkey et al. 1990). In addition to the Amadori pathway of glycation, it has been shown that metal-catalyzed oxidation of reducing sugars can give rise to ketoaldehydes, which can react with lysine residues of proteins to yield Schiff base adducts possessing a carbonyl function (Wolf and Dean 1987). For a recent review on the role of glycation in aging and disease, see Kristal and Yu (1992). For purposes of the present discussion, it is important to note that some glycation and glycoxidation intermediates contain reactive carbonyl groups and could contribute to the pool of protein carbonyls. Whether this contribution is quantitatively significant remains to be established. It is noteworthy that the glycation and glycoxidation reactions with proteins are not to be confused with the well-characterized glycosylation reaction in which carbohydrates are attached to asparagine amide or serine, threonine, and tyrosine hydroxyl groups to yield so-called glycoproteins. In contrast to the glycation and glycoxidation products, glycoproteins do not react with carbonyl reagents under the standard methods for protein carbonyl measurements (Oliver et al. 1987a; Levine et al. 1990, 1994), and therefore do not contribute to the protein carbonyl pool.

11.3 Carbonyl Group Generation as a Marker of Protein Oxidation

Because carbonyl groups are generated in the direct oxidation of amino acid residues and by reactions with lipid peroxidation products, glycation, glycoxidation, and peptide cleavage reactions, it was suggested that the carbonyl content of proteins might be used as a marker for radical-mediated protein oxidation (Oliver et al. 1984; Fucci et al. 1983; Stadtman 1988a, 1992). In the meantime, this proposition has gained support from results of numerous studies showing that the level of protein carbonyls is enhanced under conditions of oxidative stress, in various pathological disorders, and in aging.

11.3.1 Protein Oxidation in Oxidative Stress

Results of studies showing that exposure of animals or cell cultures to various conditions of oxidative stress leads to the generation of carbonyl derivatives of proteins are summarized in Table 1.

Results of these studies show that hyperoxia, X-irradiation, too much exercise, cigarette smoke, activation of neutrophils, ischemia-reperfusion, magnesium deficiency, and exposure to paraquat are among conditions of oxidative stress that can provoke oxidative damage to proteins.

11.3.2 Protein Oxidation in Disease

The oxidative modification of proteins as assessed by the presence of protein carbonyl groups is associated with a number of pathological processes, including rheumatoid arthritis (Chapman et al. 1989), muscu-

Table 1. Generation of protein carbonyls by oxidative stress

Stress condition	Tissue or cells analyzed	References
Hyperoxia	Rat hepatocytes	Starke-Reed and Oliver (1989)
	Houseflies	Sohal et al. (1993a)
	Rat lung	Winter and Liehr (1991)
Exercise	Rat hind leg muscle	Witt et al. (1992)
	Houseflies	Sohal et al. (1993a)
	Rat skeletal muscle	Reznick et al. (1992)
Ischemia reperfusion	Gerbil brain	Oliver et al. (1990)
	Rat heart	Poston and Parenteau (1992)
	Rat lung	Ischiropoulos and Al-Mehdi (1995) Ayene et al. (1993)
Neutrophil activation	Neutrophil proteins	Oliver (1987), Karsek-Staples and Webster (1993)
Rapid correction of hyponatremia	Rat brain	Mickel et al. (1990)
Paraquat toxicity	Hamster	Winter and Liehr (1991)
Magnesium deficiency	Rat kidney, brain	Stafford et al. (1993)
Cigarette smoke	Human plasma	Reznick et al. (1992)
Xanthine/xanthine oxidase	Endothelial cells	Karsek-Staples and Webster (1992)
MCO system	Human plasma	Shacter et al. (1994)
	Rat neurofilament	Troncoso et al. (1995)
Ozone	Human plasma	Cross et al. (1992)
	Rat heart	Kelly and Birch (1993)
Peroxynitrite	Rat lung	Ischiropoulos and Al-Mehdi (1995)
X-irradiation	Housefly flight muscle mitochondria	Sohal and Dubey (1994)
	Heart and brain of mice	Sohal et al. (1993b)

lar dystrophy (Murphy and Keher 1989), amyotrophic lateral sclerosis (Bowling et al. 1993), Alzheimer's disease (Smith et al. 1991, 1992; Carney et al. 1994; Harris et al. 1995; Chauhan et al. 1991), cataractogenesis (Garland et al. 1988), respiratory distress syndrome (Gladstone and Levine 1994), and probably cystic fibrosis (Brown and Kelly 1994), and progeria and Werner's syndrome (Oliver et al. 1987a). Protein oxidation is also associated with atherosclerosis (Steinbrecher et al. 1987, 1991; Steinberg et al. 1989; Uchida et al. 1994), but whether protein carbonyl derivatives are among the oxidation products has not been determined.

11.3.3 Protein Oxidation in Aging

The proposition that free radical-mediated oxidation of proteins is involved in aging (Fucci et al. 1983; Oliver et al. 1984) is based on four lines of evidence:

1. During aging, a number of enzymes accumulate as catalytically inactive or less active, heat labile forms (Dreyfus et al. 1978; Rothstein 1977, 1984).
2. Protein modifications producing changes similar to these age-related alterations can be provoked in vivo by brief exposure of animals to oxidative stress (Starke et al. 1987; Starke-Reed and Oliver 1989) and in vitro by exposing purified enzymes to active oxygen-generating systems (Fucci et al. 1983; Oliver et al. 1987b).
3. The intracellular level of oxidized protein, as determined by the protein carbonyl content, increases exponentially as a function of age (Fig. 4). Thus, the carbonyl content of proteins in cultured human fibroblasts increases exponentially as a function of the age of the fibroblast donor (Oliver et al. 1987a); the carbonyl content of proteins in the occipital lobe of human brain (Smith et al. 1991), the human eye lens cortex (Garland et al. 1988), rat hepatocytes (Starke-Reed and Oliver 1989), whole body protein of the housefly (Sohal et al. 1993a), and brain and kidney of mice (Sohal et al. 1994) increases with age.
4. Physiological conditions or treatments that extend the life span lead to a decrease in the intracellular level of protein carbonyls and vice

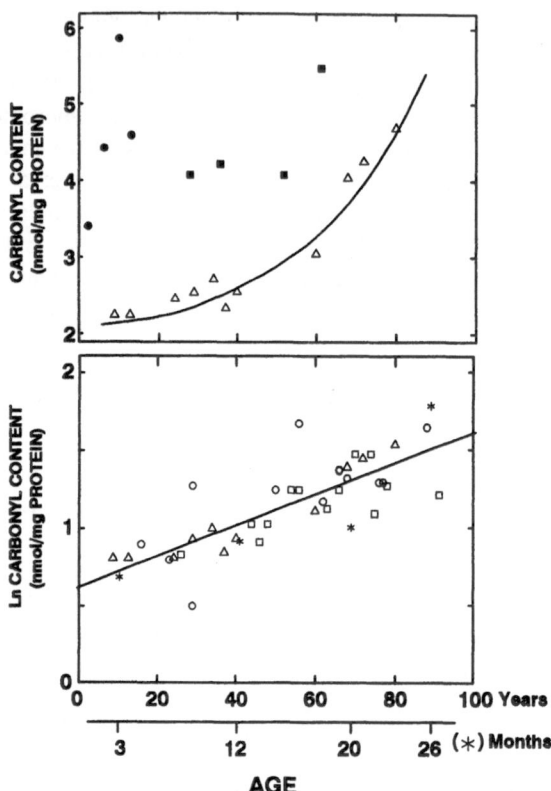

Fig. 4. Age-related accumulation of oxidized protein. Above, carbonyl content of cultured dermal fibroblasts from normal individuals (*open triangles*), from patients with progeria (*filled circles*), from patients with Werner's syndrome (*filled squares*), adapted from Oliver et al. (1987a). Below, semilog plots of the carbonyl content of proteins versus age for: dermal fibroblasts from normal individuals (*open triangles*), a replot of data *above*; the occipital lobe of the human brain tissue (*open circles*), replot of data from Smith et al. (1991); the human eye lens cortex (*open squares*), replot of data from Garland et al. (1988); rat liver hepatocytes (*), replot of data from Starke-Reed and Oliver (1989). The scale from 0 to 26 months refers to the rat hepatocytes. The scale from 0 to 100 refer to data for human subjects

versa. Thus, when examined at the same chronological age, the protein carbonyl content of a group of short-lived houseflies is higher than in the long-lived cohorts (Sohal et al. 1993a); the protein carbonyl content of brain and heart tissue of the house mouse, which has a maximum life span potential (MLSP) of 3.5 days, is considerably higher than the carbonyl contents of these tissues in the closely related white-footed mouse (MLSP = 8 days). Furthermore, the white-footed mouse is more resistant to protein damage by X-irradiation than the house mouse. These characteristics correlate well with the fact that the mitochondria of the house mouse release more H_2O_2 during respiration than that released by the white-footed mouse (Shoal et al. 1993b); dietary restriction of rats leads to an increase in the life span and also increases the resistance of the animals to X-irradiation (Youngman et al. 1992). Diet restriction in mice leads to an extension of life span, prolongation of mortality rate doubling time, and a decrease in the levels of protein carbonyl in the brain, heart, and kidney, relative to changes in ad libitum-fed animals (Sohal et al. 1994).

During aging, the resistance of animals to oxidative protein damage elicited by hyperoxia declines (Starke et al. 1987); exposure of old houseflies to X-irradiation leads to a greater increase in protein carbonyl content in the flight muscle mitochondria than is obtained with young flies (Sohal and Dubey 1994). The age-related decrease in resistance could be due either to a loss of antioxidant defenses, an increase in the rate of ROS generation, or to an age-related conversion of proteins to forms that are more susceptible to oxidative damage. Perhaps the failure to find consistent correlations between aging and one or another of these parameters (Swartz and Mader 1995) is because a change in any one parameter might lead to a manifestation that is indistinguishable from changes elicited by either one of the others. Therefore, the overall effect of an age-related alteration of a given parameter in one individual could be the same as that provoked by alteration of a different parameter in another individual. In addition to carbonyl group formation, a role of ROS in protein damage during aging is indicated by age-related changes of other markers of protein oxidation, such as the oxidation of histidine residues (Gordillo et al. 1988); formation of dityrosine residues and dityrosine-like fluorescent protein derivatives (Wells-

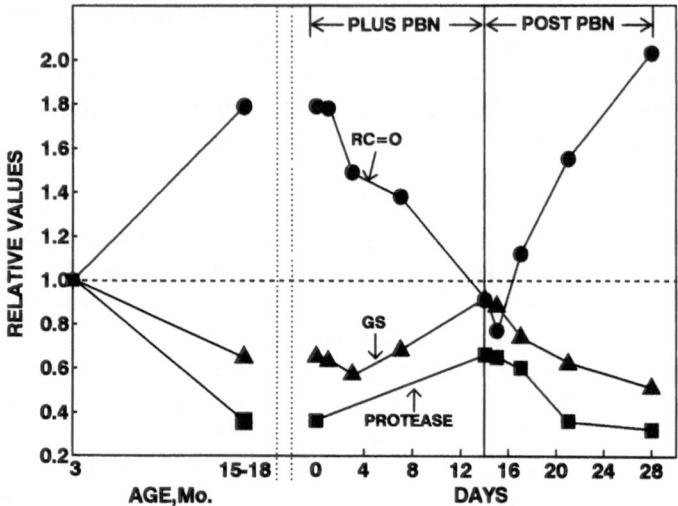

Fig. 5. Reversal of age-related changes. *Left panel*, protein carbonyl content (*filled circle*), glutamine synthetase activity (*filled triangle*), and neutral protease activity of gerbil brains as a function of animal age (filled square). *Middle panel*, time-dependent changes during chronic treatment of old gerbils (twice daily) with 35 mg of PBN per kilogram of body weight. *Right panel*, time-dependent changes that occur after the PBN treatment was discontinued

Knecht et al. 1993); oxidation of sulfhydryl groups (Zhou and Gafni 1991); and age-associated conversion of enzymes to more heat-labile, less catalytically active forms (Rothstein 1977), similar to those provoked in vitro by treatment of purified enzymes with MCO and other radical-generating systems (Oliver et al. 1987b; Takahashi and Goto 1990; Szweda and Stadtman 1992).

11.3.3.1 Reversal of Age-Related Changes

The importance of free radicals in the age-related changes in carbonyl content and accumulation of altered forms of some enzymes is highlighted by the studies of Carney et al. (1991) showing that chronic administration of a free radical spin-trap , phenyl-butyl nitrone (PBN), to "old" gerbils leads to a reversal of some age-related biochemical markers in the brain to values equal to those found in young gerbils

(Fig. 5). This was associated with a gain in temporal spatial memory function. Significantly, when the administration of PBN was discontinued, all of the biochemical markers of aging returned to values characteristic of the "old" animal.

11.3.3.2 Molecular Basis of Oxidized Protein Accumulation During Aging

Two observations suggest that the accumulation of oxidized protein during aging is a result of age-associated genetic alterations:

1. The level of oxidized protein in cultured fibroblasts is determined by the age of the donor and not by the number of cell-doublings in culture, at least over the intermediate range of cell passages.
2. The reversal of age-related changes by the administration of PBN is maintained only under conditions of chronic PBN treatment. When the treatment is discontinued, the aging markers all returned to old animal values. These observations suggest that high levels of oxidized proteins found in tissues of old animals reflect age-related changes in one or more of a multiplicity of genetic factors that control either the rate of ROS formation or the factors that govern either the susceptibility of proteins to oxidation or removal (proteolytic degradation) of the oxidized proteins (Oliver et al. 1981; Levine et al. 1981; Stadtman 1986, 1988b; Pacifici et al. 1989; Giulivi and Davies 1993).

11.4 Alzheimer's Disease

Alzheimer's disease (AD) is a common neurodegenerative disorder, the manifestation of which includes loss of memory, speech, cognitive function, and behavioral activity. AD increases exponentially as a function of age and is accompanied by a loss of synapses, neurons, overproduction of the amyloid precursor protein (APP), an increase in the intracellular Ca^{2+}, altered phosphorylation of brain enzymes/proteins, and loss or abnormal distribution (compartmentalization) of several enzymes. For review, see Katzman and Saitoh (1991), Harmon (1993), Hensley et al. (1995), and Harrington and Wischik (1995). The histopathological characteristics of AD are the presence of neuritic plaques

(NP) composed of aggregated β-amyloid protein fragments together with numerous proteins and neurofibrillary tangles (NFT) composed of paired helical filaments containing tau protein, ubiquitin, and probably microtubule-associated protein 2 (MAP 2) (Katzman and Saitoh 1991). There is considerable evidence that reactive oxygen species and β-amyloid-derived peptide radicals are involved in the development of NP and NFT.

A role of oxidative protein damage in the development of AD was suggested by the studies showing that during aging there is a progressive increase in the level of protein carbonyls and concomitant decrease in the levels of glutamine synthetase in the brain of normal individuals and in patients suffering from Alzheimer's disease (Smith et al. 1991, 1992; Carney et al. 1994). The possibility that abnormal processing of APP is involved in the development of neuritic plaques is supported by studies showing that APP can give rise to a 40–43 amino acid peptide fragment, referred to as the amyloid β protein (Aβ), which is derived from the intramembrane domain of APP (Kang et al. 1987). The amyloid protein is cytotoxic to cultured neurons (Behl et al. 1994) and is present in neuritic plaques in an aggregated form (reviewed by Behl et al. 1994; Hensley et al. 1995). Participation of reactive oxygen species in Aβ-provoked cytotoxicity is supported by the results of Behl et al. (1994) showing that: Aβ causes an increase in the cellular level of H_2O_2; antioxidants protect neuronal cells from Aβ-induced toxicity; Aβ induces formation of NF-$_κ$B, a transcription factor that is under metabolic control by oxidative stress factors; the Aβ-induced production of H_2O_2 accumulation is blocked by inhibitors of flavin oxidases, suggesting that Aβ activates a member of this class of enzyme that is known to generate ROS; clones of neuronal cells selected for resistance to Aβ-toxicity are also resistant to H_2O_2 toxicity; and catalase prevents Aβ-provoked toxicity. The oxygen free radical theory of AD pathology is supported also by results of in vitro studies of AD model systems using a synthetic 40 amino acid peptide Aβ-(1–40), possessing the same amino acid sequence as endogenous Aβ. Addition of this peptide to aqueous solution leads to slow fragmentation and the generation of peptide radicals as determined by electron paramagnetic resonance measurements and reactive oxygen species (probably OH•) as determined by the hydroxylation of salicylic acid (Hensley et al. 1994a). Moreover, a synthetic 11-amino acid peptide, Aβ-(25–35), corresponding to the C-terminal sequence of Aβ-(1–40), promotes rapid production of free radicals and ROS, and initiates synaptosomal lipoprotein oxidation, and causes inactivation of glutamine synthetase (GS) and creatine kinase (CK) (Hensley et al. 1994a,b). The activities of GS and CK are lower in AD brains than in normal control subjects (Oliver et al. 1990; Carney et al. 1994). There is a positive correlation between the regional distribution of NP density in human brain and the concentration of protein carbonyls. Compared to normal brains the amyloid-rich regions (inferior parietal lobe and the hippocampus) contain higher concentrations of protein carbonyls and also of proteins with altered infrastructure as measured by the W/S ratio (Hensley et al. 1995). The W/S ratio is defined as the ratio of the amplitudes of the weak/strong electron spin resonance (EPR) spectral lines obtained when the spin-trap 2,2,6,6-tetramethyl-4-maleimidopiperidine-1-oxyl (MAL-6) binds to sulfhydryl groups of membrane proteins. Changes in the W/S ratio are attributable to changes in the protein infrastructure (Hensley et al. 1994a). The W/S ratio of MAL-6-labeled rodent neocortical synaptosomal membrane proteins decreases as a function of animal age and also under conditions of oxidative stress, including iron loading, hyperoxia ischemia-reperfusion trauma, and cortical X-irradiation (reviewed by Hensley et al. 1994a). Significantly, the protein carbonyl content of various AD brain regions parallels the regional pattern of W/S ratio values (Hensley et al. 1995).

11.4.1 Modification of Glutamine Synthetase

A potential role of glutamine synthetase (GS) in neurotoxicity derives from the consideration that GS plays a key role in brain metabolism. It catalyzes the removal of toxic ammonia and converts the neurotransmitter and excitotoxic amino acid glutamate to the non-neurotoxic glutamine. The possibility that loss of GS activity contributes to AD is indicated by the observations:

1. The activity of GS decreased in various regions of the brain during aging and AD (Smith et al. 1991; Carney et al. 1994).
2. GS is present in the cerebrospinal fluid of patients suffering from Alzheimer's disease (Gunnerson and Haley 1992).

3. Aβ fragments increase glutamate excitotoxicity (Yankner et al. 1990.
4. GS enhances the neurotoxicity of β-amyloid peptide and is converted by these peptides to a catalytically inactive form (Aksenov et al. 1995).

Based on the fact that Aβ is neurotoxic in vitro, it was proposed (Hensley et al. 1994b) that abnormal proteolytic conversion of the β-amyloid protein precursor protein leads to the release of various peptide fragments such as Aβ-(1–40) into the extraneuronal space. These fragments may then undergo further fragmentation to form toxic peptide radicals and ROS, which can attack cell membranes, initiating lipoperoxidation and damage to membrane proteins, and thus compromise ion homeostasis and facilitate Ca^{2+} influx. The Aβ-derived radicals could also attack enzymes to generate covalently bonded aggregates and, among others, damage glutamate transporters and thereby potentiate glutamate excitotoxicity.

In summary, there is abundant evidence that protein oxidation is associated with aging, AD, and several other diseases. In the case of AD, there is growing evidence that abnormal processing of the β-amyloid precursor protein leads to the production of fragments that can give rise to peptide radicals and ROS that damage enzymes/proteins and membrane lipids, and thereby contribute to the formation of plaques and neurofibrillary tangles that are characteristic of Alzheimer's disease.

References

Aksenov MY, Aksenova MV, Harris ME, Hensley K, Butterfield DA, Carney JM (1995) Enhancement of β-amyloid peptide Aβ(1–40)-mediated neurotoxicity by glutamine synthetase. J Neurochem 65:1899–1902

Allen DR, Wallis GL, McCay PB (1994) Catechol adrenergic agents enhance hydroxyl radical generation in xanthine oxidase systems containing ferritin: implications for ischemia/reperfusion. Arch Biochem Biophys 315:235–243

Alonso E, Cervera J, Garcia-Espan A, Bendala E, Rubio V (1992) Oxidative inactivation of carbonyl phosphate synthetase (ammonia). J Biol Chem 267:4524–4532

Amici A, Levine RL, Tsai L, Stadtman ER (1989) Conversion of amino acid residues in proteins and amino acid homopolymers to carbonyl derivatives by metal-catalyzed oxidation reactions. J Biol Chem 264:3341–3346

Ayene IS, Al-Medi AB, Fisher AB (1993) Inhibition of lung tissue oxidation during ischemia/reperfusion by 2-mercaptopropionyl glycine. Arch Biochem Biophys 303:307–312

Beckman JS, Beckman TW, Chen J, Marshall PA, Freeman B (1990) Apparent hydroxyl radical production by peroxynitrite: implications for endothelial injury from nitric oxide and superoxide. Proc Natl Acad Sci USA 87:1620–1624

Behl C, Davis JB, Lesley R, Schubert D (1994) Hydrogen peroxide mediates amyloid β protein toxicity. Cell 77:817–827

Berlett BS, Omar OHM, Sahakian JA, Levine RL, Stadtman ER (1991) Oxidation of proteins by ozone. FASEB J 5:A6691

Berlett BS, Yim MB, Chock PB, Stadtman ER (1995) Peroxynitrite-provoked nitration of tyrosine residues in glutamine synthetase causes changes in regulatory properties similar to those obtained by ATP-dependent adenylylation. FASEB J 9:A1287

Bovaris A, Oshino N, Chance B (1972) The cellular production of hydrogen peroxide. Biochem J 128:617–630

Bowling AC, Schultz JB, Brown RH Jr, Beal MF (1993) Superoxide dismutase activity, oxidative damage, and mitochondrial energy metabolism in familial and sporadic amyotrophic lateral sclerosis. J Neurochem 61:2322–2325

Brown RK, Kelly F (1994) Evidence for increased oxidative damage in patients with cystic fibrosis. Pediatr Res 36:487–493

Brunori M, Rotilio G (1984) Biochemistry of oxygen radical species. Methods Enzymol 105:22–35

Cairo G, Tacchini L, Pogliaghi G, Anzon E, Tomasi A, Bernelli-Zazzera A (1995) Induction of ferritin synthesis by oxidative stress. Transcriptional and posttranscriptional regulation by expansion of the free iron pool. J Biol Chem 270:700–703

Carney JM, Starke-Reed PE, Oliver CN, Landum RW, Cheng MS, Wu JF, Floyd RA (1991) Reversal of age-related increase in brain protein oxidation, decrease in enzyme activity loss, and loss of temporal and spatial memory by chronic administration of the spin-trapping compound, N-tert-butyl-α-phenylnitrone. Proc Natl Acad Sci USA 88:3633–3636

Carney JM, Smith CD, Carney AM, Butterfield A (1994) Aging- and oxygen-induced modifications in brain biochemistry and behavior. In: Franceschi C, Crepaldi G, Cristofalo VJ (eds) Aging and cellular defense mechanisms. N Y Acad Sci 63:110–119

Chapman ML, Rubin BR, Gracy RW (1989) Increased carbonyl content of proteins in synovial fluid from patients with rheumatoid arthritis. J Rheumatol 16:15–19

Chauhan A, Chauhan VPS, Brockerhoff H, Wisniewski HM (1991) Action of amyloid β-protein on protein kinase C activity. Life Sci 49:1555–1562

Chevion M (1988) A site-specific mechanism for free radical induced biological damage: the essential role of redox-active transition metals. Free Radic Biol Med 5:27–37

Climent I, Tsai L, Levine RL (1989) Derivatization of γ-glutamyl semialdehyde residues in oxidized proteins by fluoresceinamine. Anal Biochem 182:226–232

Creeth JM, Cooper B, Donald ASR, Clamp JR (1983) Studies of the limited degradation of mucus glycoproteins. Biochem J 211:323–332

Cross CE, Reznick AZ, Packer L, Davis PA, Suzuki YJ. Halliwell B (1992) Oxidative damage to human plasma proteins by ozone.Free Radic Res Commun 15:347–352

Dreyfus JC, Kahn A, Schapira F (1978) Posttranscriptional modifications of enzymes. Curr Top Cell Regul 14:243–297

Esterbauer H, Schaur RJ, Zolner H (1991) Chemistry and biochemistry of 4-hydroxynonenal, malondialdehyde, and related aldehydes. Free Radic Biol Med 11:81–128

Farber JM, Levine RL (1986) Sequence of a peptide susceptible to mixed-function oxidation: probable cation binding site in glutamine synthetase. J Biol Chem 261:4575–4578

Friguet B, Stadtman ER, Szweda LI (1994) Modification of glucose-6-phosphate dehydrogenase by 4-hydroxy-2-nonenal. J Biol Chem 269:21639–21643

Fucci L, Oliver CN, Coon MJ, Stadtman ER (1983) Inactivation of key metabolic enzymes by mixed-function oxidation reactions: possible implication in protein turnover and aging. Proc Natl Acad Sci USA 80:1521–1525

Garland D, Russell P, Zigler JS (1988) Oxidative modification of lens proteins. In: Simic MG, Taylor KS, Ward JF, von Sontag V (eds) Oxygen radicals in biology and medicine. Plenum, New York, pp 347–353

Garrison WM (1987) Reaction mechanisms in the radiolysis of peptide, polypeptides, and proteins. Chem Rev 87:381–398

Garrison WM, Jayko ME, Bennett W (1962) Radiation-induced oxidation of proteins in aqueous solution. Radiat Res 16:487–502

Giulivi C, Davies KJA (1993) Dityrosine and tyrosine oxidation products are endogenous markers for selective proteolysis of oxidatively modified red blood cell hemoglobin by (the 19S) proteasome. J Biol Chem 268:8752–8759

Gladstone IM Jr, Levine RL (1994) Oxidation of proteins in neonatal lungs. Pediatrics 93:764–768

Gordillo E, Ayala A, F-Lobato M, Bautista J, Machado A (1988) Possible involvement of histidine residues in the loss of activity of rat liver malic enzyme during aging. J Biol Chem 263:8053–8057

Grimes HD, Perkins KK, Boss WF (1983) Ozone degrades into hydroxyl radical under physiological conditions. Plant Physiol 72:1016–1020

Gunnersen D, Haley B (1992) Detection of glutamine synthetase in cerebrospinal fluid of Alzheimer's disease patients: a potential biochemical marker. Proc Natl Acad Sci USA 89:11949–11953

Halliwell B, Gutteridge JMC (1989) Free radicals in biology and medicine. Clarendon, Oxford

Harmon D (1993) Free radical theory of aging: a hypothesis on pathogenesis of senile dementia of the Alzheimer's type. Age 16:23–30

Harrington CR, Wischick CM (1995) Pathogenic mechanisms in Alzheimer syndromes and related disorders. In: Esser K, Martin GM (eds) Molecular aspects of aging. Wiley, Chichester, pp 227–239

Harris M, Hensley K, Butterfield DA, Leedle RA, Carney JM (1995) Direct evidence of oxidative injury produced by Alzheimer's β-amyloid peptide (1–40) in cultured hippocampal neurons. Exp Neurol 131:193–202

Hensley K, Carney JM, Hall N, Shaw W, Butterfield DA (1994a) Electron paramagnetic resonance investigations of free radical-induced alterations in neocortical synaptosomal membrane protein infrastructure. Free Radic Biol Med 17:321–331

Hensley K, Carney JM, Mattson MP, Askenova M, Harris M, Wu JF, Floyd RA, Butterfield DA (1994b) A model for β-amyloid aggregation and neurotoxicity based on free radical generation by the peptide: relevance to Alzheimer's disease. Proc Natl Acad Sci USA 91:3270–3274

Hensley K, Hall N, Subramanian R, Cole P, Harris M, Askenova M, Gabbita P, Wu JF, Carney JM, Lovell M, Marksbery WR, Butterfield DA (1995) Brain regional correspondence between Alzheimer's disease histopathology and biomarkers of protein oxidation. J Neurochem 65:2146–2156

Huggins TG, Wells-Knecht MC, Detorie NA, Baynes JW, Thorpe SR (1993) Formation of o-tyrosine and dityrosine in proteins during radiolytic and metal-catalyzed oxidation. J Biol Chem 268:12341–12347

Ischiropoulos H, Al-Mehdi AB (1995) Peroxynitrite-mediated oxidative protein modifications. FEBS Lett 364:279–282

Kang J, Lemaire H-G, Unterbeck A, Salbaum JM, Masters CL, Grzeschik KH, Multhaup G, Beyreuther K, Muller-Hill B (1987) The precursor of Alzheimer's disease amyloid A4 protein resembles a cell surface receptor. Nature 325:733–736

Katzman R, Saitoh T (1991) Advances in Alzheimer's disease. FASEB J 5:278–286

Kelley FJ, Birch S (1993) Ozone exposure-inhibits cardiac protein sythesis in the mouse. Free Radic Biol Med 14:443–446

Kim K, Rhee SG, Stadtman ER (1985) Nonenzymatic cleavage of proteins by reactive oxygen species generated by dithiothreitol and iron. J Biol Chem 260:15394–15397

Kristal BS, Yu, BP (1992) An emerging hypothesis: synergistic induction of aging by free radicals and Maillard reactions. J Gerontol 47:B107–B114

Krsek-Staples JA, Webster RO (1993) Ceruloplasmin inhibits carbonyl formation in endogeneous cell proteins. Free Radic Biol Med 14:115–125

Lamb DJ, Leake DS (1994) Iron released from ferritin at acidic pH can catalyze the oxidation of low density lipoprotein. FEBS Lett 352:15–18

Levine RL (1983) Oxidation of glutamine synthetase. II. Characterization of the ascorbate model system. J Biol Chem 258:11828–11833

Levine RL, Oliver CN, Fulks RM, Stadtman ER (1981) Turnover of bacterial glutamine synthetase: oxidative modification precedes proteolysis. Proc Natl Acad Sci USA 78:2120–2124

Levine RL, Garland D, Oliver CN, Amici A, Climent I, Lenz AG, Ahn B-W, Shaltiel S, Stadtman ER (1990) Determination of carbonyl groups in oxidatively modified proteins. Methods Enzymol 186:464–478

Levine RL, Williams JA, Stadtman ER, Schacter E (1994) Carbonyl assays for determination of oxidatively modified proteins. Methods Enzymol 233:346–357

Mickel HS, Oliver CN, Starke-Reed PE (1990) Protein oxidation in magnesium deficient rat brains and kidneys. Biochem Biophys Res Commun 196:92–97

Monnier V (1990) The Maillard reaction and the aging process. J Gerontology 45:B105–B111

Monnier VM, Gerhardinger C, Marion MS, Taneda S (1995) Novel approaches toward inhibition of the Maillard reaction in vivo: search, isolation, and characterization of prokaryotic enzymes which degrade glycated substrates. In: Cutter RG, Packer L, Bertram J, Mori A (eds) Oxidative stress and aging. Birkhäuser, Basel, pp 141–149

Mullarkey CJ, Edelstein D, Brownlee M (1990) Free radical generation by early glycation products: a mechanism for accelerated atherogenesis in diabetes. Biochem Biophys Res Commun 173:932–939

Murphy ME, Kehrer JP (1989) Oxidation state of tissue thiol groups and content of protein carbonyl groups in chickens with inherited muscular dystrophy. Biochem J 260:359–364

Neuzil J, Gebiki JM, Stocker R (1993) Radical-induced chain oxidation of proteins and its inhibition by chain-breaking antioxidants. Biochem J 293:601–606

Oliver CN (1987) Inactivation of enzymes and oxidative modification of proteins by stimulated neutrophils. Arch Biochem Biophys 253:62–72

Oliver CN, Levine RL, Stadtman ER (1981) Regulation of glutamine synthetase turnover. In: Holzer H (ed) Metabolic interconversion of enzymes. Springer, Berlin Heidelberg New York, pp 259–258

Oliver C, Fucci L, Levine R, Wittenberger M, Stadtman ER (1982) Inactivation of key metabolic enzymes by P450 linked mixed function oxidation systems. In: Hietanen E, Laitinen M, Hanninen O (eds) Cytochrome P-450, biochemistry, biophysics, and environmental implications. Elsevier Biomedical, Amsterdam, pp 531–538

Oliver CN, Fulks R, Levine RL, Fucci L, Rivett AJ, Roseman JE, Stadtman ER (1984) Oxidative inactivation of key metabolic enzymes during aging. In: Roy AK, Chatterjee B (eds) Molecular basis of aging. Academic, New York, pp 235–262

Oliver CN, Ahn B-W, Moreman EJ, Goldstein S, Stadtman ER (1987a) Age-related changes in oxidized proteins. J Biol Chem 262:5488–5491

Oliver CN, Levine RL, Stadtman ER (1987b) A role of mixed-function oxidation reactions in the accumulation of altered enzyme forms during aging. JAGS 35:947–956

Oliver CN, Starke-Reed PE, Stadtman ER, Liu GJ, Carney JM, Floyd RA (1990) Oxidative damage to brain proteins, loss of glutamine synthetase activity, and production of free radicals during ischemia/reperfusion-induced injury to gerbil brain. Proc Natl Acad Sci USA 87:5144–5147

Pacifici RE, Salo DC, Davies KJA (1989) Macroxyproteinase (MOP): a 70 kDA proteinase complex that degrades oxidatively denatured proteins in red blood cells. Free Radic Biol Med 7:521–536

Palinski W, Rosenfeld ME, Yla-Herttuala S, Gurtner GC, Socher SS, Butler SW, Parthasarathy S, Carew TE, Steinberg D, Witzum JL (1989) Low density lipoprotein undergoes oxidative modification in vivo. Proc Natl Acad Sci USA 86:1372–1376

Poston JM (1988) Detection of oxidized amino acid residues using P-amino benzoic acid adducts. Fed Proc 46:1979 (abstract)

Poston JM, Parenteau GL (1992) Biochemical effects of ischemia on isolated perfused rat heart tissues. Arch Biochem Biophys 295:35–41

Pryor WA (1994) Mechanisms of radical formation from reactions of ozone with target molecules in the lung. Free Radic Biol Med 17:451–465

Pryor WA, Jin X, Squadrito GL (1994) One- and two-electron oxidations of methionine by peroxynitrite. Proc Natl Acad Sci USA 91:11173–11177

Reznick AZ, Witt E, Matsumoto M, Packer L (1992) Vitamin E inhibits protein oxidation in skeletal muscle of resting and exercised rats. Biochem Biophys Res Commun 189:801–806

Rothstein M (1977) Recent developments in age-related alterations of enzymes. Mech Aging Dev 6:241–257

Rothstein M (1984) Changes in enzymatic proteins during aging. In: Roy AK, Chatterjee B (eds) Molecular basis of aging. Academic, New York, pp 209–232

Schauenstein E, Esterbauer H (1979) Formation and properties of reactive aldehydes. In: Submolecular biology of cancer. CIBA Found Ser 67:225–224

Shacter E, Williams JA, Lim M, Levine RL (1994) Differential susceptibility of plasma proteins to oxidative modification: examination by Western blot immunoassay. Free Radic Biol Med 17:429–437

Smith CD, Carney JM, Starke-Reed PE, Oliver CN, Stadtman ER, Floyd RA (1991) Excess brain protein oxidation and enzyme dysfunction in normal and Alzheimer's disease. Proc Natl Acad Sci USA 88:10540–10543

Smith CD, Carney JM, Tatsuno T, Stadtman ER, Floyd RA, Markesbery WR (1992) Protein oxidation in aging brain. In: Franceschi C, Crepaldi G, Cristofalo VG, Vijg J (eds) Aging and cellular defense mechanisms. N Y Acad Sci 663:110–119

Sohal RS, Dubey A (1994) Mitochondrial oxidative damage, hydrogen peroxide release, and aging. Free Radic Biol Med 16:621–626

Sohal RS, Agarwal S, Dubey A, Orr WC (1993a) Protein oxidative damage is associated with life expectancy of houseflies. Proc Natl Acad Sci USA 90:7255–7259

Sohal RS, Ku H-H, Agarwal S (1993b) Biochemical correlates of longevity in two closely related rodent species. Biochem Biophys Res Commun 196:7–11

Sohal RS, Ku H-H, Agarwal S, Forster MJ, Lal H (1994) Oxidative damage, mitochondrial oxidant generation and antioxidant defenses during aging and in response to food restriction in the mouse. Mech Aging Dev 74:121–133

Stadtman ER (1986) Oxidation of proteins by mixed-function oxidation systems: implications in protein turnover, aging, and neutrophil function. Trends Biochem Sci 11:11–12

Stadtman ER (1988a) Biochemical markers of aging. Exp Gerontol 23:327–347

Stadtman ER (1988b) Protein modification in aging. J Gerontol 43:B112–B120

Stadtman ER (1990) Metal ion-catalyzed oxidation of proteins: biochemical mechanism and biological consequences. Free Radic Biol Med 9:315–325

Stadtman ER (1992) Protein oxidation and aging. Science 257:1220–1224

Stadtman ER (1995) The status of oxidatively modified proteins as a marker of aging. In: Esser K, Martin GM (eds) Molecular aspects of aging. Wiley, Chichester, pp 130–143

Stadtman ER, Oliver CN (1991) Metal catalyzed oxidation of proteins: physiological consequences. J Biol Chem 266:2005–2008

Stadtman ER, Wittenberger ME (1985) Inactivation of *Escherichia coli* glutamine synthetase by xanthine oxidase, nicotinate hydroxylase, horseradish peroxidase, or glucose oxidase: effect of ferredoxin putaredoxin, and menadione. Arch Biochem Biophys 239:379–387

Stafford RE, Mak TM, Kramer JH, Weglicki WB (1993) Protein oxidation in magnesium deficient rat brains and kidneys. Biochem Biophys Res Commun 196:596–600

Starke-Reed PE, Oliver CN (1989) Protein oxidation and proteolysis during aging and oxidative stress. Arch Biochem Biophys 275:559–567

Starke PE, Oliver CN, Stadtman ER (1987) Modification of hepatic proteins in rats exposed to high oxygen concentration. FASEB J 1:36–39

Steinberg D, Parthasarathy S, Carew TE, Koo JC, Witzum JW (1989) Beyond cholesterol modifications of low-density lipoprotein that increase artherogenicity. N Engl J Med 320:915–924

Steinbrecher UP, Witzum JL, Parthasarathy S, Steinberg D (1987) Decrease in reactive amino groups during oxidation or endothelial cell modification of LDL. Correlation with changes in receptor-stimulated catabolism. Atherosclerosis 7:135–143

Steinbrecher UP, Zang H, Lougheed M (1991) Role of oxidatively modified LDL in atherosclerosis. Free Radic Biol Med 9:155–168

Swallow AJ (1960) Effect of ionizing radiation on proteins. RCO groups, peptide bond cleavage, inactivation, -SH oxidation. In: Swallow AJ (ed) Radiation chemistry of organic compounds. Pergamon, New York, pp 211–224

Swartz HM, Mader K (1995) Free radicals in aging: theories, facts, and artifacts. In: Esser K, Martin GM (eds) Molecular aspects of aging. Wiley, New York, pp 78–97

Szweda LI, Stadtman ER (1992) Iron catalyzed oxidative modification of glucose-6-phosphate dehydrogenase from Leuconostoc mesenteroides. J Biol Chem 267:3096–3100

Taborsky G (1973) Oxidative modification of proteins in the presence of ferrous iron and air. Effect of ionic constituents of the reaction medium on the nature of the oxidation products. Biochemistry 12:1341–1348

Takahashi R, Goto S (1990) Alteration of aminoacyl-tRNA synthetase with age: heat labilization of the enzyme by oxidative damage. Arch Biochem Biophys 277:228–233

Toyokuni S, Uchida K, Okamoto K, Hattori-Nakakuki Y, Hiai H, Stadtman ER (1994) Formation of 4-hydroxy-2-nonenal-modified proteins in the renal proximal tubules of rats treated with a renal carcinogen, ferric nitrilotriacetate. Proc Natl Acad Sci USA 91:2616–2620

Troncoso JC, Costello AC, Kim JH, Johnson GVW (1995) Metal catalyzed oxidation of bovine neurofilaments in vitro. Free Radic Res Biol Med 18:891–899

Uchida K, Kawakishi S (1993) 2-oxohistidine as a novel biological marker for oxidatively modified proteins. FEBS Lett 332:208–210

Uchida K, Kawakishi S (1994) Identification of oxidized histidine generated at the active site of Cu,Zn-superoxide dismutase exposed to H_2O_2. J Biol Chem 269:2405–2410

Uchida K, Stadtman ER (1993) Covalent modification of 4-hydroxynonenal to glyceraldehyde-3-phosphate dehydrogenase. J Biol Chem 268:6388–6393

Uchida K, Kato Y, Kawakishi S (1990) A novel mechanism for oxidative cleavage of prolyl peptides induced by the hydroxyl radical. Biochem Biophys Res Commun 169:265–271

Uchida K, Toyokuni S, Nishikawa K, Kawakishi S, Oda H, Hiai H, Stadtman ER (1994) Michael addition-type 4-hydroxy-2-nonenal adducts in modified low-density lipoproteins: markers for atherosclerosis. Biochemistry 33:12487–12494

Vogt W (1995) Oxidation of methionyl residues in proteins: Tools, targets, and reversal. Free Radic Biol Med 18:93–105

Wells-Knecht MC, Huggins TG, Dyer G, Thorpe SR, Baynes JW (1993) Oxidized amino acids in lens protein with age. J Biol Chem 268:12348–12352

Winter ML, Liehr JG (1991) Free radical-induced carbonyl content in protein of estrogen-treated hamsters assayed by sodium boro[^{3}H]hydride reduction. J Biol Chem 266:14446–14450

Witt EH, ReznickAZ, Viguie CA, Starke-Reed PE, Packer L (1992) Exercise, oxidative damage, and effects of antioxidant manipulation. J Nutr 122:766–773

Wolf SP, Dean RT (1987) Glucose autooxidation and protein modification. Biochem J 245:243–250

Yankner BA, Duffy LK, Kirschner DA (1990) Neurotropic and neurotoxic effects of amyloid protein: reversal by tachykiniv neuropeptides. Science 250:279–282

Youngman LD, Park J-YK, Ames B (1992) Protein oxidation associated with aging is reduced by dietary restriction of protein calories. Proc Natl Acad Sci USA 89:9112–9116

Zhou JQ, Gafni A (1991) Exposure of rat muscle phosphoglycerate kinase to a nonenzymatic MFO system generates the old form of the enzyme. J Gerontol 46:B217–B221

Subject Index

Ernst Schering Research Foundation Workshop

Editors: Günter Stock
Ursula-F. Habenicht